Power Systems

Electrical power has been the technological foundation of industrial societies for many years. Although the systems designed to provide and apply electrical energy have reached a high degree of maturity, unforeseen problems are constantly encountered, necessitating the design of more efficient and reliable systems based on novel technologies. The book series Power Systems is aimed at providing detailed, accurate and sound technical information about these new developments in electrical power engineering. It includes topics on power generation, storage and transmission as well as electrical machines. The monographs and advanced textbooks in this series address researchers, lecturers, industrial engineers and senior students in electrical engineering.

Power Systems is indexed in Scopus

More information about this series at https://link.springer.com/bookseries/4622

Mehdi Rahmani-Andebili
Editor

Applications of Artificial Intelligence in Planning and Operation of Smart Grids

Editor
Mehdi Rahmani-Andebili
Electrical Engineering Department
Montana Technological University
Butte, MT, USA

ISSN 1612-1287 ISSN 1860-4676 (electronic)
Power Systems
ISBN 978-3-030-94524-4 ISBN 978-3-030-94522-0 (eBook)
https://doi.org/10.1007/978-3-030-94522-0

This Springer imprint is published by the registered company Springer Nature Switzerland AG
The registered company address is: Gewerbestrasse 11, 6330 Cham, Switzerland

Preface

Artificial intelligence (AI) is going to play a significant role in smart grid planning and operation, especially in solving its real-time problems, as it is fast, adaptive, robust, and less-dependent on power system's accurate model and parameters. Machine learning, as a current application of AI, can get data from its environment and learn to deal with it. Generally, machine learning is categorized in three different types including supervised learning, unsupervised learning, and reinforcement learning.

This book covers the recent research advancements in some of the applications of AI in the planning and operation of smart grids. It applies AI-based techniques to deal with the smart grid issues, presents the innovative approaches, and studies the numerical problems.

The first chapter of the book presents a new market modeling approach based on an agent-based machine learning technique to improve the efficiency of the power market while considering the technical aspects of the power system.

The second chapter applies reinforcement learning technique to reach the maximum power point tracking (MPPT) of a photovoltaic (PV) system to increase its efficiency under different temperature and solar irradiance conditions.

In the third chapter, a hybrid model, including neighborhood component analysis (NCA) and ANN, is applied for the short-term power prediction of a PV system.

The fourth chapter studies different statistical and deep learning algorithms in both short-term and long-term power consumption predictions.

Butte, MT, USA Mehdi Rahmani-Andebili

Contents

About the Editor

Mehdi Rahmani-Andebili is an assistant professor in the Electrical Engineering Department at Montana Technological University, MT, USA. Before that, he was also an assistant professor in the Engineering Technology Department at State University of New York, Buffalo State, NY, USA, during 2019–2021. He received his first MSc and PhD degrees in electrical engineering (power system) from Tarbiat Modares University and Clemson University in 2011 and 2016, respectively, and his second MSc degree in physics and astronomy from the University of Alabama in Huntsville in 2019. Moreover, he was a postdoctoral fellow at Sharif University of Technology during 2016–2017. As a professor, he has taught many courses such as Essentials of Electrical Engineering Technology, Electrical Circuits Analysis I, Electrical Circuits Analysis II, Electrical Circuits and Devices, Industrial Electronics, Renewable Distributed Generation and Storage, Feedback Controls, DC and AC Electric Machines, and Power System Analysis. Dr. Rahmani-Andebili has more than 200 single-author and first-author publications including journal papers, conference papers, textbooks, books, and book chapters. He is an IEEE Senior Member and the permanent reviewer of many credible journals. His research areas include smart grid, power system operation and planning, integration of renewables and energy storages into power system, energy scheduling and demand-side management, plug-in electric vehicles, distributed generation, and advanced optimization techniques in power system studies.

Chapter 1
A New Agent-Based Machine Learning Strategic Electricity Market Modelling Approach Towards Efficient Smart Grid Operation

P. Kiran, K. R. M. Vijaya Chandrakala, S. Balamurugan, T. N. P. Nambiar, and Mehdi Rahmani-Andebili

Abstract The electricity market operation in smart grid requires a certain level of intelligence by the agents. The operation of multiple agents in the market is different under various scenarios where the machine learning approach exhibited by the agents plays a major role. Each trader is focused on achieving high profit yielding and a positive reward by experimenting various cost parameters and propensity levels. The best learning model for smart grid incorporating the deregulated market structure is reinforcement learning. In this type of model, the agent takes action in an environment based on the previous experience and improvises automatically in the successive bidding environment. This help the generator companies to maximize their profit and achieve smart bidding environment for efficient smart grid transactions. Once the generator fails to meet the requirement and propose greater price, penalization also becomes a part to assess the bidding strategy in electricity market through agent-based machine learning. A new interactive Variant Roth-Erev algorithm based approach is implemented in various agents, and the test system considered for the entire analysis is an IEEE 5-bus system. The system is modelled taking into account different propensity levels and stopping rules. Congestion handling in the system is also an objective irrespective of earning high profit. Further, the analysis is extended for a period of 50 days to showcase the effectiveness of the

P. Kiran · K. R. M. Vijaya Chandrakala (✉) · S. Balamurugan · T. N. P. Nambiar
Department of Electrical and Electronics Engineering, Amrita School of Engineering, Coimbatore, Amrita Vishwa Vidyapeetham, India
e-mail: cb.en.d.eee15002@cb.students.amrita.edu; krm_vijaya@cb.amrita.edu; s_balamurugan@cb.amrita.edu; tnp_nambiar@amrita.edu

M. Rahmani-Andebili
Electrical Engineering Department, Montana Technological University, Butte, MT, US
e-mail: mrahmaniandebili@mtech.edu

M. Rahmani-Andebili (ed.), *Applications of Artificial Intelligence in Planning and Operation of Smart Grids*, Power Systems,
https://doi.org/10.1007/978-3-030-94522-0_1

approach. The results obtained for different cases are investigated in detailed way through comparison without learning and with machine learning approach using Java-based programming.

Keywords Deregulated Electricity Market · Smart Grid · Locational marginal Price (LMP) · DC Optimal Power Flow (DCOPF) · Reinforcement Learning · Interactive Variant Roth-Erev (VRE) algorithm

1.1 Introduction

Deregulated electricity market introduces competition in energy market among the traders, with open access of infrastructure for improved reliability of service and high efficiency. For many decades, since their inception, electric power industries have monopolized the way in which they are operated and sold electricity to customers within their geographical region. This vertically integrated utility is obligated to sell electric power and meet the needs of its customers. These are publicly owned and not operated for profit, and the tariffs are set by the regulatory organizations. The reforms in power sector have introduced new competitive markets where there is open access of system infrastructure. In deregulated power system, generating companies (GenCos) are owned by different power suppliers and they compete to sell energy to the consuming utilities at competitive prices. Transmission companies (TransCos) will transmit power to the load centres over high voltage lines. Distribution companies (DisCos) will distribute power at the wholesale or retail level to the customers. The Independent system operator (ISO) will make sure the reliability and fair transmission line services along with the scheduling and dispatching of power in real time scenario [1, 2]. According to Electricity Act, 2003, of India, one of the main stages of reformation is the division of state-owned electricity boards into different entities like Generating Companies, Transmission companies and Distribution Companies. The whole electricity market is controlled by an ISO and evaluates the Market Clearing Price (MCP) according to the bids and offers submitted to the market. Even though there is sufficient generation, system may experience congestion in transmission line, making it difficult for power delivery. In order to manage the congestion in transmission line, a pricing scheme is introduced called as Locational Marginal Pricing (LMP), which depends on the timing and location of injection of power (MW). The market operates on a day-ahead basis considering different bids and offers along with the real time operation where there is high probability for congestion happening in transmission lines. In such situations, the deregulated electricity market having many agents will make use of LMP [3–6]. Even though the introduction of competition in the power market enhances the reliability and security in the power sector, certain key issues needs more clarification. This includes (i) competition which could effectively substitute for regulation in some segments related to energy trading and settlements, (ii) necessity of best approach required for the power sector reform for the given size and structure which may vary from country to country, (iii) involvement of smaller power sectors and (iv) challenges related to mitigation of market power. These factors motivated the need for an efficient electricity market structure having the interactive learning capability and managing the congestion through LMP structure [7–9].

Agent-based GenCos require an advanced learning algorithm based on reinforcement learning technique. The optimum scheduling of the generator to meet the demand is exercised through economic scheduling of the generators for profit maximization. In order to achieve this, a new interactive learning algorithm which configures the GenCo as an agent in AMES is necessary. In most of the previously published works, different supervised and unsupervised techniques have been applied [10–12]. In this chapter, a unique algorithm is developed, namely, interactive Variant Roth-Erev (VRE) reinforcement technique, which is analysed and implemented on wholesale day-ahead electricity market. The chapter focuses on the dynamic action in the deregulated electricity market entities as agents and its optimal scheduling in economic point of visualization. The work also focuses on the development of a new interactive learning algorithm and configures the GenCo agent on the day-ahead wholesale electricity market operation. The chapter is briefed as follows: In Sect. 1.2, the transmission pricing using LMP and various market simulators are explained. Section 1.3 deals with agent-based model of electricity market and its dynamic actions. Different machine learning techniques applicable for smart grid operation are described in Sect. 1.4. The section also introduces a new interactive VRE algorithm applicable for electricity market for taking intelligent actions. Section 1.5 explains in detail about the implementation of agent-based electricity market and different analysis undertaken. Section 1.6 concludes the work.

1.2 Pricing Mechanism in Electricity Market and Market Simulators

1.2.1 Locational Marginal Price (LMP)

Transmission pricing scheme provides new cost of transmission system to the consumers partially or fully. One of the commonly used pricing technique was Megawatt-Mile method. The Megawatt-Mile pricing was computed by finding the product of the transmission line distance and power flow value. Many Latin American and Asian countries adopt this method for transmission pricing. Earlier United States and India also followed this method. Adding all the power flow-mile gives us the overall transmission price total, which gives the measure of grids used in every transaction. Megawatt-Mile pricing concept supports the effective utilization of the transmission facility. The congestion pricing component was not accurate and, hence, various congestion pricing methods were introduced. One of the problems experienced in power system operation is the congestion in power lines, which may occur in different locations in different time. The pricing of power has to take into account congestion also. Locational Marginal Price (LMP) finds the buyer's price related to competitive market energy cleared at suitable locations including the congestion cost related to the difference in LMPs between two locations [3]. This method takes into account the congestion cost and allocates the same to the one who was responsible for creating that particular congestion. The technique was on the basis of the price of catering energy to the next load rise at a specific location on the transmission network. LMP considers the congestion cost and includes a congestion component in its total pricing. LMP is also called as nodal pricing, since it is calculated at all nodes based on the bids provided by the traders to the power exchange [6]. To overcome the pricing issues due to congestion and for congestion management, LMP can be applicable. In LMP analysis, only the LSE who is creating the congestion need to pay the congestion cost irrespective of the zone. Equation 1.1 shows the calculation of the LMP at bus '*i*':

$$\lambda(i) = \lambda_{\text{ref}} + \lambda_{\text{loss}(i)} + \lambda_{\text{congestion}(i)} \tag{1.1}$$

Here, Marginal Cost due to reference bus, losses and transmission congestion are denoted as 'λ_{ref}', '$\lambda_{\text{loss}}(i)$' and '$\lambda_{\text{congestion}}(i)$' respectively.

For better understanding of the analysis, a 2-bus system is considered with no losses as referred in Fig. 1.1.

The supply bid and demand offer for the above system is highlighted in Fig. 1.2.

The negative MW means withdrawal from the system.

The optimization rule is to minimize total system cost such that total supply equals the total demand, that is Supply cost = 10*Supply MW

Demand Offer = 50 * (−Demand MW)

Optimizing the total system cost, Min (10*Supply MW + 50 * −Demand MW)

Such that the optimization meets the equality constraints, that is Supply MW is equal to Demand MW

Supply MW = Demand MW = 100 MW. Therefore, Total System Cost is, (10 * 100 + 50 * −100) = −4000 Rs/hr

To find out LMP, add one additional MW to load bus. This can be done by two ways. One way of solving is as follows: 1 MW is served by the generator. In this case, the total system cost will be −3990 Rs/hr. The second way of solving is follows: 1 MW is to reduce the load bid by 1 MW and maintaining the generation at the same dispatch level. In this case, the total system cost will be −3950 Rs/hr. Therefore, the total system cost is minimum in case of the first method 1. Therefore, relating to method 1, the change in cost = −3990 – (−4000) = Rs 10/hr.

Therefore, LMP for load bus is Rs 10/hr divided by 1 MW, that is Rs 10/ MWh.

Fig. 1.1 Two bus system

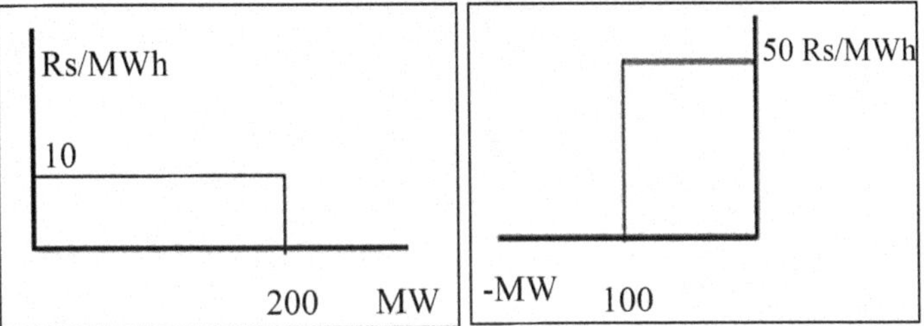

Fig. 1.2 Supply bid and demand offer

1.2.2 Electricity Market Simulators

Deregulated Electricity Market needs intelligent acting decision makers to help various operators for successful restructuring. The requirement of good simulation platform is necessary for modelling efficient electricity market. Complex decision making is a tedious task for already existing Supervisory Control and Data Acquisition Systems (SCADA) due to its lack of stability and slowness. Local decision making and implementation is difficult when the control is centralized. Multi-Agent Systems (MAS) gives fine outputs for such problems. In a market-based system, every entity needs assured amount of intelligence. MAS consists of various agents and deals with interaction of agents, its organization and target attainment. Flexibility, scalability, autonomy and reduction in problem complexity are some of the key elements of MAS [11]. A feasible approach for managing complex distributed systems is MAS. In MAS, a complex control problem is split into several small problems, in which each small problem will be solved by an individual agent as shown in Fig. 1.3.

Introduction of competition will result in reduced cost and will ensure reliable operation in both wholesale and retail electricity market. The market operator checks the balancing between the supply and the demand. The various agents like generation agents, transmission agents and distribution agents must communicate through a platform following proper standards [12]. Various poolco and bilateral transactions must be effective with respect to the electricity market. Various electricity market simulators for modelling and simulation are given in next section.

1. *Electricity Market Complex Adaptive Systems Model (EMCAS)* [13]

   This model describes the behaviour of generation companies and customers. This model is capable of calculating the price taking care of timing and location. But this model takes into account various zones and which makes the model complex [13].
2. *Multi-agent Intelligent Simulator (MAIS)* [14]

   The dynamic changes happening in wholesale electricity markets can be analysed by using this simulator. Here the agents are equipped with limited learning capabilities and hence, the agent action may not be complete [14].

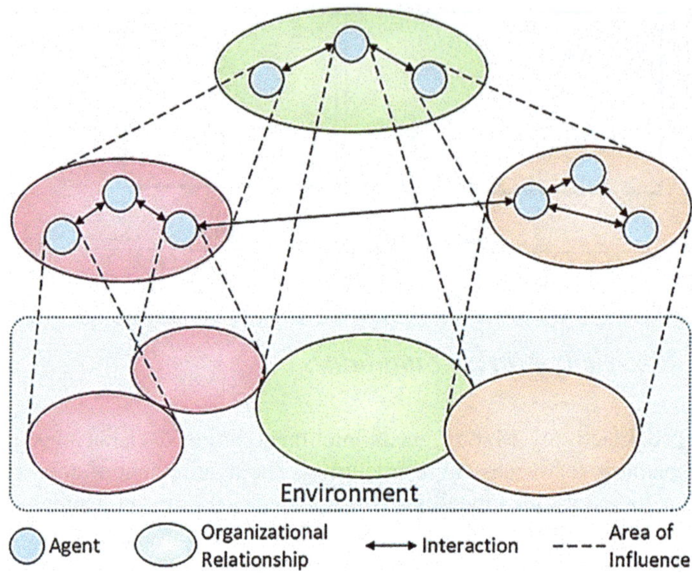

Fig. 1.3 Multi-agent system

3. *Multi-agent Simulation System for Competitive Electricity Markets (MASCEM)* [15]

   MASCEM uses reinforcement learning algorithm which consists of various agents. This can be used as a decision support tool in the operation of real time electricity markets. The architecture of MASCEM is complex, and the various virtual power players make this a commercial tool [15].
4. *Power Trading Agent Competition (Power TAC)* [16]

   This is a commercial tool and is applicable only for retail electricity markets. More level of experimentation is possible by using this simulator and is applicable for future electricity markets also. This simulator is suitable for complex computation purposes but is regarded as a business tool [16].
5. *Agent based Modelling of Electricity Systems (AMES)*

   AMES wholesale electricity market will operate in AC transmission system and can handle congestion cost also. The main advantage of traders in AMES is its learning capability. This modelling technique uses LMP for the operation in the day-ahead electricity market. The optimal power flow method used is DC Optimal Power Flow (DCOPF) and is calculated based on the bids and offers submitted to the power exchange. Next section explains about modelling an electricity market.

1.3 Agent-Based Model of Electricity Market

The major elements in an AMES electricity market are GenCo, TransCo and DisCo, and those are governed by the Independent System Operator (ISO). There are two ways by which an electricity market runs, which are either on a day ahead basis or on hour ahead basis. The AMES uses Java-enabled Reinforcement Learning Algorithm (JReLM) and that allows an incorporated learning across the GenCos. Figure 1.4 shows the AMES framework.

The Java language is used to build up the AMES as it enhances the usage and readability. The AMES architecture in Java permits the user to adapt the code to their requirements. AMES integrated with DCOPF is used for LMP simulation solved using linear programming as it substantiates the capability of the application towards speed and robustness [17–21]. The AMES ISO fixes the DC-OPF problems by the application of Java DC Optimal Power Flow (DCOPFJ) as shown in Fig. 1.5. The dynamic flow in AMES is illustrated in Fig. 1.6, where ‘m’ represents the supply offer, ‘a’ represents the actual cost and ‘R’ represents the reported cost [22].

The AMES GenCos are independent energy agents with strategic learning capabilities. The AMES tool allows the user to determine the number of GenCos and their position at various buses. Here, the GenCos can sell power only to the Load Serving Entities (LSE). The GenCos real power capacity limit is given in Eq. 1.2.

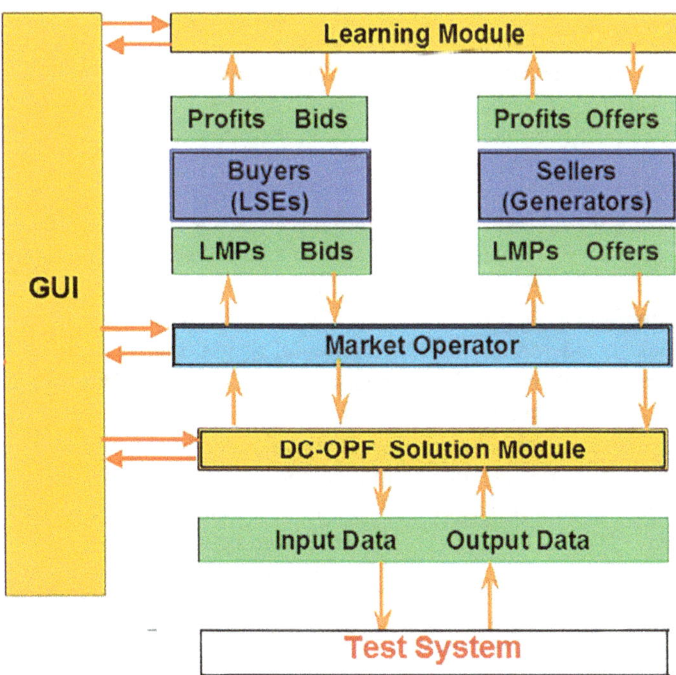

Fig. 1.4 AMES framework

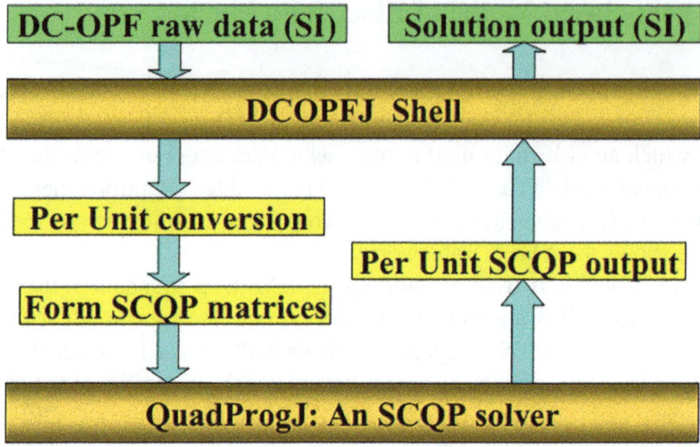

Fig. 1.5 DCOPF solver

$$\mathrm{Cap}_i^{\mathrm{L}} \;''\; \mathrm{PGi} \;''\; \mathrm{Cap}_i^{\mathrm{U}}, \tag{1.2}$$

where the lower and upper production limits are depicted as $\mathrm{Cap}_i^{\mathrm{L}}$ and $\mathrm{Cap}_i^{\mathrm{U}}$ and the real power production is PGi. Equation 1.3 gives the total cost of GenCo '*i*'.

$$\mathrm{TCi} = \mathrm{ai.PGi} + \mathrm{bi.PGi}^2 + \mathrm{FCOSTi}, \tag{1.3}$$

where ai, bi, FCOSTi are constants. Equation 1.4 gives the marginal cost of GenCo '*i*' for $i = 1$ to n

$$\mathrm{MCi} = \mathrm{ai} + 2.\mathrm{bi.PGi} \tag{1.4}$$

In the varying deregulated system, the dynamic analysis will be provided by the AMES. The analysis initiates through choosing an offer and later by updating it. In between these two actions, power flow control, transmission congestion, commitments by the traders and the market pricing have to be carried out. The various assumptions considered for the construction of action domain for the GenCos are the following: (i) the marginal cost reported will have forward sloping ($\mathrm{bi}^{\mathrm{R}} > 0$), (ii) the reported production interval is such that $\mathrm{Cap}_i^{\mathrm{RL}} < \mathrm{Cap}_i^{\mathrm{RU}}$, (iii) the true marginal cost curve must lie on or below the reported marginal cost curve and (iv) the GenCo reports its true lower production limit and reports a lower or equal true upper production limit. Equation 1.5 shows the net earnings calculation of GenCo '*i*' for hour '*H*' of a day '*D*'. Net earnings are the difference between the total cost and revenues. GenCo '*i*' net earnings for a whole day '*D*' is calculated using the Eq. 1.6.

$$\mathrm{NEi}(H,D) = \mathrm{LMP.PGi} - \mathrm{TCi}(\mathrm{PGi}) \tag{1.5}$$

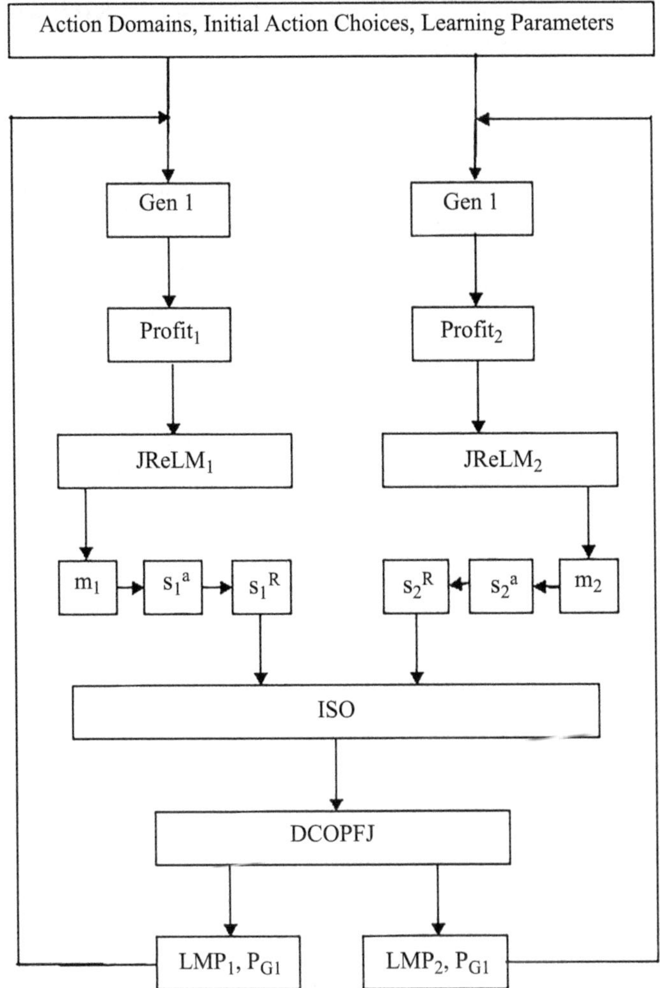

Fig. 1.6 Dynamic flow in AMES

$$\mathrm{NEi}(D) = \sum_{H=0}^{H=23} \mathrm{NEi}(H,D). \tag{1.6}$$

In the next section, different machine learning techniques applied for smart operation and the developed new interactive VRE machine learning approach are discussed.

1.4 Machine Learning Technique for Smart Electricity Market Operation

In order to make the grid smart, agents will make decision depending on its learning strategy following the sequence of action. Depending upon the past performance, the agent will learn and take actions. An agent can learn and change on the basis of its previous performances and varying conditions. Supervised or unsupervised learning algorithms are basic two techniques in machine learning and artificial intelligence. The algorithm which uses input data and labelled output data (target) is called as supervised learning. The algorithm tries to map the input to the target in a manner that generalizes well to unseen input data. Algorithms using unsupervised learning attempt to find structure in unlabelled data. The terms supervised and unsupervised do not describe well the mechanisms of reinforcement learning algorithms. Reinforcement learning is a reward-based learning technique where agent learns through interaction with the environment. The agent's goal is not to use labelled data in some sense or explicitly finding general structures in the data. As a result, reinforcement learning is considered a category of its own. The agent and the environment with which it is interacting are fundamental to any reinforcement learning algorithm. An agent is the decision maker that observes a state and decides what action to take. The set of possible actions and states are respectively called the action space 'A' and state space 'S'. The goal at each time 't' is to maximize the rewards in the future. Model-free reinforcement learning requires no model or information about the dynamics in the environment. Model-free algorithms can be divided into two subcategories as main classes of reinforcement learning: value based and policy based. The first subcategory of model-free algorithm is called value-based methods, where the approach is to approximate the action-value function 'Q', which is profit maximization under GenCo in electricity market and which helps to take an action. Policy-based methods directly parameterize the policy function, that is reward 'π' without involving the action-value function 'Q', that is profit maximization of the GenCo in the electricity market blindly in the decision-making. Game theory approach is suitable for decision making in such scenarios. So, to handle the dynamic environment of the electricity market, a model free algorithm with policy-based method could be the right choice towards taking optimum decisions by the GenCos. Game theory deals with finding some equilibrium levels, considered as a set of strategies. Reinforcement learning is emerged from game theory concept. This technique is suitable to take optimal decision for each agent to implement necessary actions. One kind of reinforcement learning technique is Q-learning that works by learning an action-value function. This does not require a model to compare the available actions. The action results in certain rewards and the main aim of agent is to maximize the reward. In reinforcement learning, Roth-Erev learning algorithm is suitable to handle model free, policy-based method in deregulated electricity market which makes the entities action as agents.

1.4.1 Interactive Variant Roth-Erev Reinforcement Learning (VRE-RL)

The initial propensity value '$q(0)$' will be equal in Roth-Erev RL algorithm for all agent actions. The action propensity is then updated based on the action selected for all time 't'. The propensity for action at time $t + 1$ is calculated using Eqs. 1.7 and 1.8.

$$q_j(t+1) = (1-r)q_j(t) + \pi_k(t)(1-e); \text{if } j = k. \tag{1.7}$$

$$q_j(t+1) = (1-r)q_j(t) + \pi_k(t)\frac{e}{N-1}; \text{if } j \neq k. \tag{1.8}$$

Here 'r' is the recency parameter, 'e' is the experimentation parameter and $\pi k(t)$ is the reward obtained for action 'k' at time 't'. The probability of choosing the action aj at time 't' is given in Eq. 1.9:

$$P_j(t) = \frac{q_j(t)}{\sum_{i=0}^{N-1} q_i(t)}. \tag{1.9}$$

The disadvantage of Roth-Erev RL algorithm is that it fails to learn for some actions whose reward is zero or negative. Therefore, it is modified slightly and in Modified Roth-Erev RL algorithm, the action propensity is slightly modified and is as given in Eq. 1.10:

$$q_j(t+1) = (1-r)q_j(t) + q_j(t)\frac{e}{N-1}; \text{if } j \neq k. \tag{1.10}$$

In certain cases, if the reward is negative, MRE algorithm fails to calculate the probability. VRE RL algorithm introduces a cooling parameter (T) which allows accurate probability calculation even if the difference between the propensities is less. In VRE RL algorithm, the choice probability is calculated using Gibbs-Boltzmann distribution making use of the exponential operator as shown in Eq. 1.11.

$$P_j(t) = \frac{e^{q_j(t)/T}}{\sum_{i=0}^{N-1} e^{q_j(t)/T}}. \tag{1.11}$$

In this chapter, a poolco model with day-ahead market operation is considered and the entities are treated as agents. The intelligence required for the agent is given through a learning strategy, and reinforcement algorithm is found to be the best. Agent-based Roth-Erev Reinforcement learning approach is the basic learning algorithm and from which a new modified interactive Variant Roth-Erev (VRE) RL

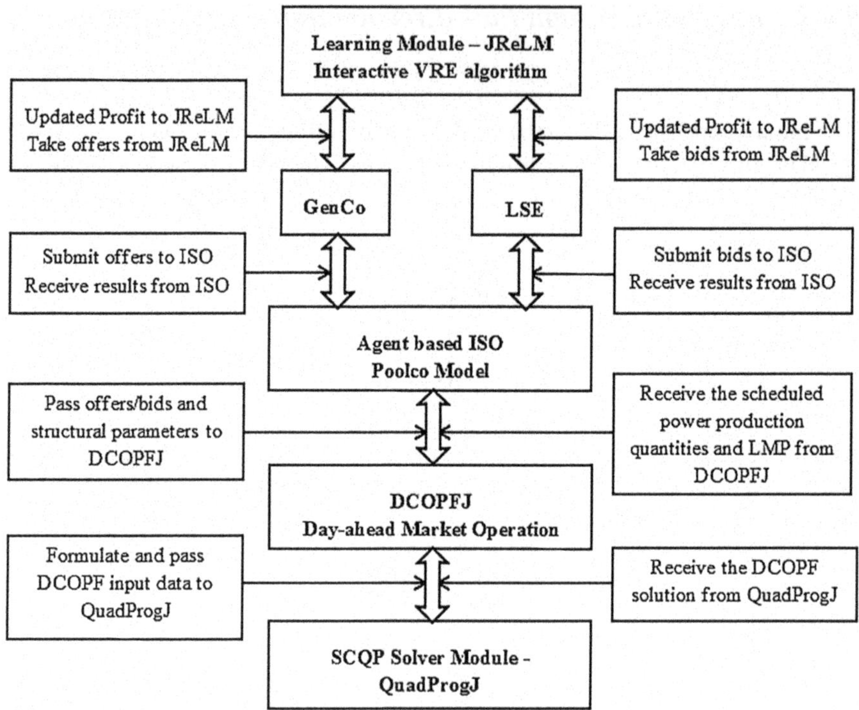

Fig. 1.7 Overview of different modules for smart market operation

algorithm is proposed in this work. For effective interaction among the GenCo agents, co-learning is exercised initially with Roth-Erev RL technique. This technique shows error when the net earnings tend to zero; hence, it is modified as Modified Roth-Erev (MRE) RL. The modified method produces good results except when the net earnings become negative. Later, the proposed method interactive VRE is developed which converges at threshold probability and achieves high net earnings. Figure 1.7 gives an overview of different modules and the smart actions taking place in a poolco day-ahead market operation with interactive VRE algorithm.

The implementation of an agent-based electricity market and analysis on diverse perspectives are explained.

1.5 Agent-Based Electricity Market Implementation and Analysis

To examine the effect of interactive VRE learning approach on GenCos as agents, a small test system is considered and is subjected for analysis with and without learning. Figure 1.8 shows an IEEE 5-bus test system where various learning and action

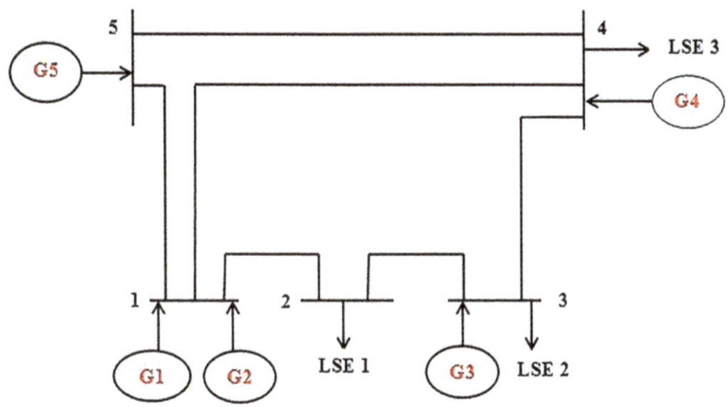

Fig. 1.8 IEEE 5-bus test system

Table 1.1 Transmission line data [17]

Line	From	To	Max. capacity (MW)	GenCo	Bid (MW)	Marginal cost ($/MW)
1	1	2	250	G1	110	14
2	1	4	150	G2	100	15
3	1	5	400	G3	520	30
4	2	3	350	G4	200	35
5	3	4	240	G5	600	10
6	4	5	240	–	–	–

Table 1.2 Cost parameters of GenCos [17]

GenCo No.	Fcost ($/h)	a ($/MWh)	b ($/MW^{2}h)	Cap^L (MW)	Cap^U (MW)
1	56.9	14	0.005	0	110
2	0.11	15	0.006	0	100
3	2267.53	25	0.01	0	520
4	5.19	30	0.012	0	200
5	1391.16	10	0.007	0	600

domain parameters are varied and analysis is conducted. Tables 1.1, 1.2, and 1.3 gives the Line data, Cost data and LSE data, respectively. LMP without learning is shown in Fig. 1.9.

Out of the five LMPs in each node, LMP 2 is high and there is wide difference among the LMP values at various nodes which indicates the presence of congestion in transmission line. Figure 1.10 shows the GenCo commitment without learning.

The GenCos learns as per the parameters given in Table 1.4 and the obtained LMP for day 1 is shown in Fig. 1.11. Comparing Figs. 1.9 and 1.11 in learning case, there is a significant increase in LMP outcomes in all buses. The GenCo commitment with learning is shown in Fig. 1.12.

Table 1.3 LSE data [17]

Hour	LSE 1 (MW)	LSE 2 (MW)	LSE 3 (MW)	Hour	LSE 1 (MW)	LSE 2 (MW)	LSE 3 (MW)
0	350	300	250	12	403.82	346.13	288.44
1	322.93	276.8	230.66	13	394.8	338.4	282
2	305.04	261.47	217.89	14	390.37	334.6	278.83
3	296.02	253.73	211.44	15	390.37	334.6	278.83
4	287.16	246.13	205.11	16	408.25	349.93	291.61
5	291.59	249.93	208.28	17	448.62	384.53	320.44
6	296.02	253.73	211.44	18	430.73	369.2	307.67
7	314.07	269.2	224.33	19	426.14	365.26	304.39
8	358.86	307.6	256.33	20	421.71	361.47	301.22
9	394.8	338.4	282	21	412.69	353.73	294.78
10	403.82	346.13	288.44	22	390.37	334.6	278.83
11	408.25	349.93	291.61	23	363.46	311.53	259.61

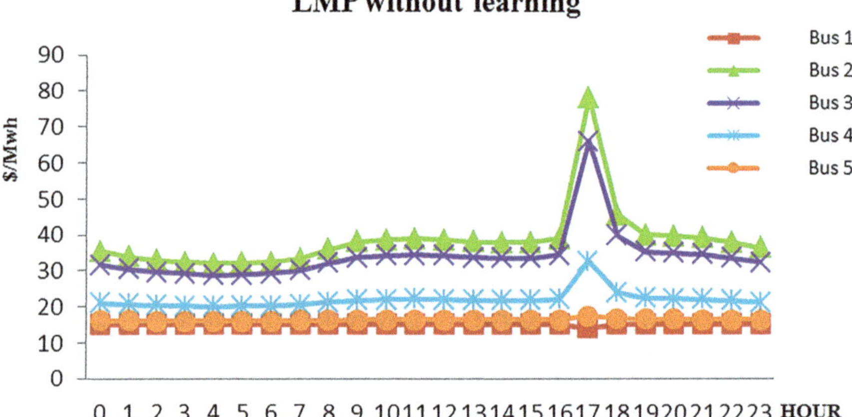

Fig. 1.9 LMP without learning

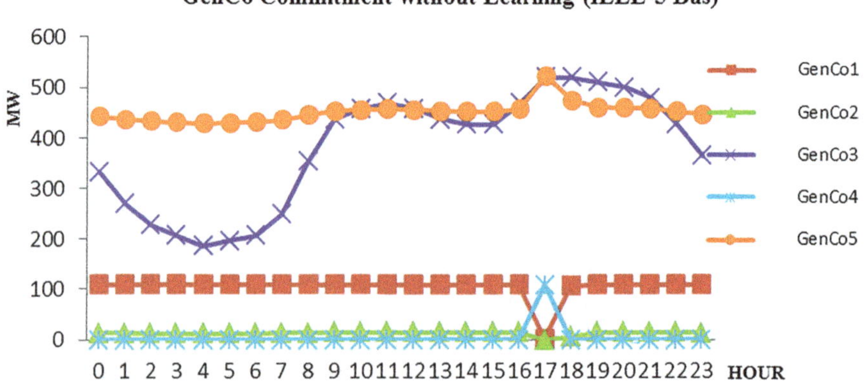

Fig. 1.10 GenCo commitment without learning

Table 1.4 Learning parameters in VRE algorithm

$q(0)$	C	r	e	$RIMax^L$	$RIMin^C$	SS
6000	1000	0.04	0.97	0.75	1	0.001

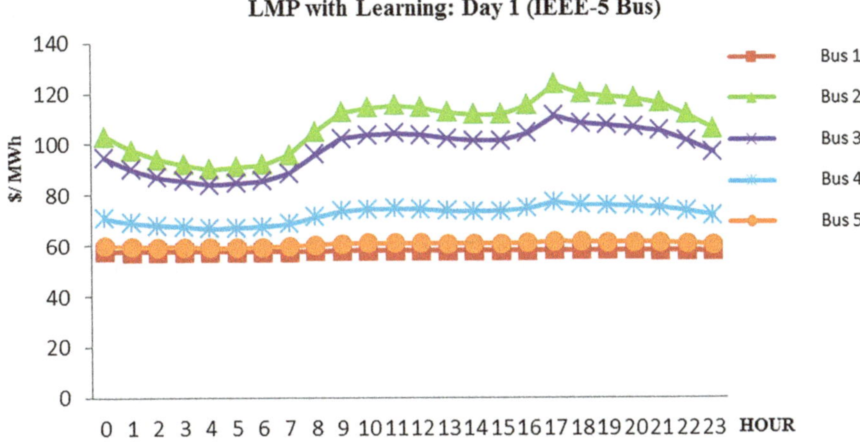

Fig. 1.11 LMP with learning

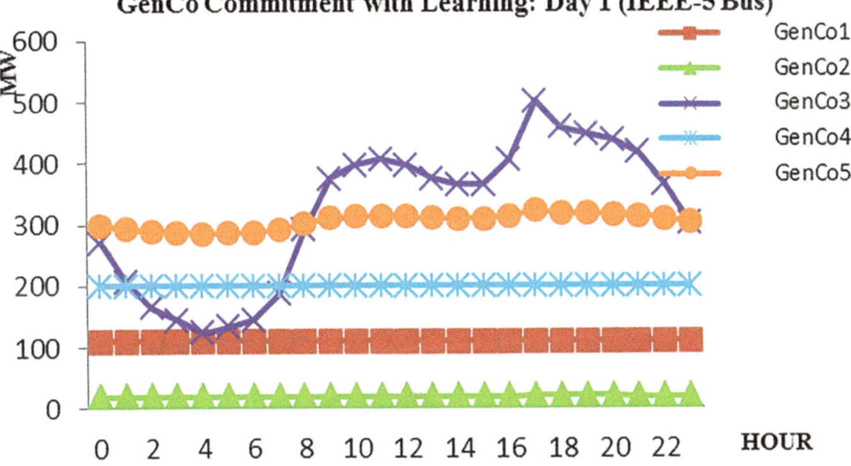

Fig. 1.12 GenCo commitment with learning

Figs. 1.11 and 1.12 clearly indicates that the high cost GenCo 4 uses its learning capability to make it dispatched in almost all hours of a day in contrast to no learning case, where it is dispatched only at hour 17. The results also show the changes happening in other GenCos also. Learning makes certain GenCo to be the winner and several others to be looser. Therefore, through reinforcement learning, the

GenCo learns the profitable strategy to report high marginal cost. This results in substantial increase in the LMP values but the effect on dispatch and power flow will be less. The GenCos learn in such a way that it reports a higher price than true marginal price over a reported operating capacity. The commitment for hour 17 with and without learning is shown in Fig. 1.13.

The GenCo commitments along with net earnings for a day are shown in Table 1.5. The net earnings of GenCo 5 and GenCo 3 are high, whereas the net earnings of the high cost GenCo 4 are low and is the less preferred GenCo.

An effective smart grid operation consists of numerous types of generation sources with distributed generation. In an extended analysis, a 10 MW distributed generation-based micro-Grid (MG) is connected to node 2, node 3 and node 4 of the test system shown in Fig. 1.8. It is modelled as a non-dispatchable source of generation. The objective is to find out the optimum location of MG in the electricity market. The LMP is calculated for all three cases and results obtained show that the optimum location in terms of economics is node 2. Table 1.6 shows the LMP values with and without MG at hour 17. Also, the effect of MG is less evident in the wholesale power market scenario as given in Fig. 1.14. The analysis shows that micro-grid integration along with learning is quite possible in the deregulated electricity market. The micro-grid comprises of sources with intermittent nature which need to be modelled in a retail electricity market for analysing its effect on the main grid. The micro-grid behaviour and learning performance will be more visible in a retail electricity market when compared to a wholesale electricity market.

Based on day 1 analysis, the experimentation parameter (e) and recency parameter (r) as mentioned in Table 1.4 with low propensity of $q = 6000$, medium propensity of $q = 60{,}000$, high propensity of $q = 160{,}000$ is considered for improved profit decision making by GenCos for smart bidding through reinforcement learning algorithm. The convergent action is chosen by using the stopping rule which is the

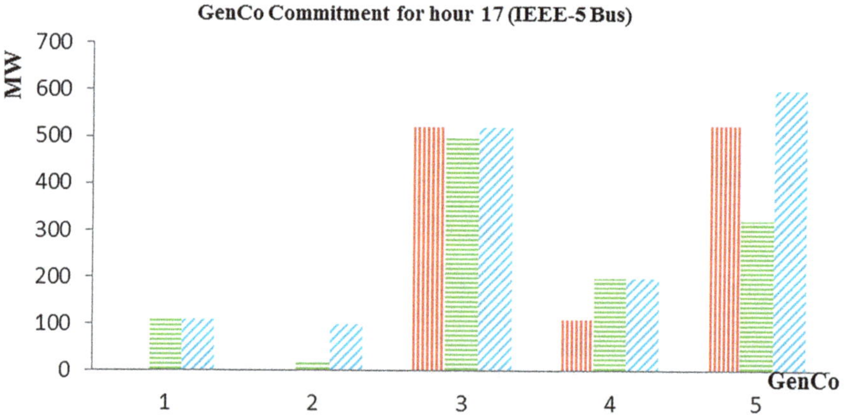

Fig. 1.13 GenCo commitment for hour 17

Table 1.5 GenCo daily net earnings

	GenCo1		GenCo2		GenCo3		GenCo4		GenCo5	
Hour	Commitment (MW)	Net earnings ($/h)	Commitment (MW)	Net earnings ($/h)	Commitment (MW)	Net earnings ($/h)	Commitment (MW)	Net earnings ($/h)	Commitment (MW)	Net earnings ($/h)
0	110	67.81	13.8726	1.15	332.534	1105.79	0	0	443.593	1377.42
1	110	67.24	13.4369	1.08	269.449	726.02	0	0	437.537	1340.07
2	110	66.87	13.1585	1.04	227.712	518.53	0	0	433.537	1315.68
3	110	66.69	13.0208	1.02	206.664	427.1	0	0	431.523	1303.48
4	110	66.49	12.8745	0.99	185.994	345.94	0	0	429.536	1291.51
5	110	66.58	12.9381	1	196.395	385.71	0	0	430.527	1297.48
6	110	66.69	13.0208	1.02	206.664	427.1	0	0	431.523	1303.48
7	110	67.05	13.2968	1.06	248.768	618.85	0	0	435.553	1327.95
8	110	68	14.0135	1.18	353.201	1247.51	0	0	445.576	1389.76
9	110	68.77	14.5999	1.28	437.017	1909.84	0	0	453.629	1440.45
10	110	68.95	14.7335	1.3	458.06	2098.19	0	0	455.64	1453.25
11	110	69.03	14.7971	1.31	468.394	2193.93	0	0	456.626	1459.55
12	110	68.95	14.7335	1.3	458.06	2098.19	0	0	455.64	1453.25
13	110	68.77	14.5999	1.28	437.017	1909.84	0	0	453.629	1440.45
14	110	68.66	14.5117	1.26	426.667	1820.44	0	0	452.622	1434.06
15	110	68.66	14.5117	1.26	426.667	1820.44	0	0	452.622	1434.06
16	110	69.03	14.7971	1.31	468.394	2193.93	0	0	456.626	1459.55
17	2.0709	0.02	0	0	520	18654.46	108.88	142.27	522.635	1912.03
18	107.34	57.61	6.11408	0.22	520	4983.82	0	0	474.153	1573.75
19	110	69.41	15.0831	1.37	510.081	2601.82	0	0	460.626	1485.24
20	110	69.32	15.0129	1.35	499.827	2498.27	0	0	459.642	1478.9
21	110	69.14	14.8807	1.33	478.747	2291.98	0	0	457.629	1465.97
22	110	68.66	14.5117	1.26	426.667	1820.44	0	0	452.622	1434.06
23	110	68.09	14.0808	1.19	363.949	1324.59	0	0	446.602	1396.17

Table 1.6 LMP values with and without MG

LMP (Hour 17)	Without MG	With MG (At node 2)	With MG (At node 3)	With MG (At node 4)
LMP 1 ($/MWh)	14.02	14.03	14.19	14.04
LMP 2 ($/MWh)	78.24	74.79	75.49	77.40
LMP 3 ($/MWh)	66.07	63.32	63.87	65.40
LMP 4 ($/MWh)	32.61	31.77	31.94	32.38
LMP 5 ($/MWh)	17.32	17.28	17.34	17.29

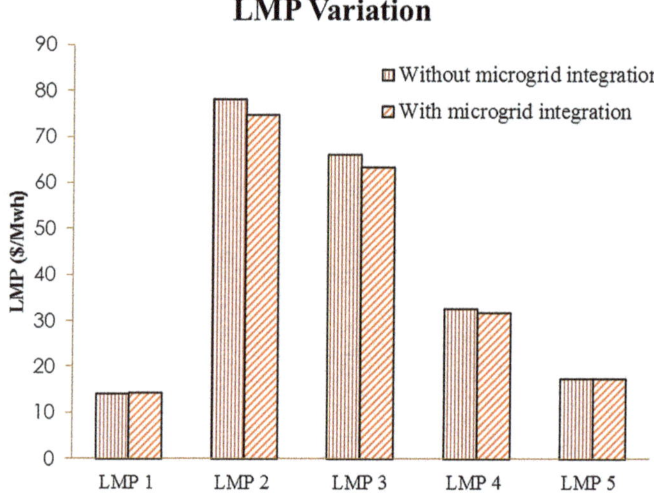

Fig. 1.14 LMP variation with and without MG

maximum day and threshold probability limits. Here, for the current analysis, a threshold probability limit of 0.999 is considered which is same for all '*n*' bus systems. The convergent with different propensity values in IEEE 5-Bus test system is shown in Table 1.7.

Table 1.7 indicates that all five GenCos after learning at day 5 with low propensity have helped to earn profit reaching the threshold probability of 0.999 when compared to medium and high propensity reaching the convergence at day 8 and day 12 for the same profit making by all the learnt GenCos to make smart bidding to the ISO using the algorithm.

This faster convergence with the choice of the parameter selection in the reinforcement algorithm helps GenCos to improve its net earnings and schedule its generation through smart bidding. GenCo 3 is highly preferred GenCo due to its high capacity. Table 1.8 gives the Daily Net Earnings (DNE) while learning is applied on GenCo 3 and shows how GenCo 3 is learnt by reporting new 'a' and 'b' cost parameters.

In the next analysis, the co-learning effect is applied initially between GenCo 1 and GenCo 3 and later between GenCo 3 and GenCo 5 and its behaviour on the

Table 1.7 IEEE-5 bus system convergent action (threshold probability of 0.999)

When $q = 6000$, Convergent Action in Day 5; When $q = 60{,}000$, Convergent Action in Day 8; When $q = 160{,}000$, Convergent Action in Day 12

$c = 1000$, $r = 0.04$ (Low), $e = 0.97$ (High)

Day	GenCo	Low initial propensity, $q = 6000$			Medium initial propensity, $q = 60{,}000$			High initial propensity, $q = 160{,}000$		
		Propensity	Probability	Profit ($/Day)	Propensity	Probability	Profit ($/Day)	Propensity	Probability	Profit ($/Day)
1	1	10323.6	0.47	114090.0	62163.6	0.35	114090.0	158163.6	0.17	114090.0
	2	6585.1	0.02	20627.6	58425.1	0.01	20627.6	154425.1	0.01	20627.6
	3	28003.4	1	556085.2	79843.4	1	556085.2	175843.4	1	556085.2
	4	13393.4	0.95	190836.9	65233.4	0.92	190836.9	161233.4	0.81	190836.9
	5	19864.3	0.99	352609.1	71704.3	0.99	352609.1	167704.3	0.99	352609.1
2	1	13830.7	0.97	98002.3	59774.6	0.16	98002.3	152865.5	0.08	98002.3
	2	9048.3	0.23	86573.0	59317.4	0.15	86573.0	152408.3	0.06	86573.0
	3	49593.1	1	567746.6	99359.5	1	567746.6	191519.5	1	567746.6
	4	19765.8	0.99	172701.7	69532.2	0.99	172701.7	161692.2	0.99	172701.7
	5	28159.5	1	227242.8	77925.9	1	227242.8	170085.9	1	227242.8
–	–	–	–	–	–	–	–	–	–	–
–	–	–	–	–	–	–	–	–	–	–
5	**1**	**23532.6**	**0.99**	**98002.3**	54849.2	0.14	98002.3	139731.2	0.08	98002.3
	2	**17984.1**	**0.99**	**86573.0**	54392.0	0.05	86573.0	139274.0	0.05	86573.0
	3	**109317.5**	**1**	**567746.6**	153071.9	1	567746.6	234553.7	1	567746.6
	4	**37393.8**	**1**	**172701.7**	81853.2	1	172701.7	163477.1	1	172701.7
	5	**51106.6**	**1**	**227242.8**	98009.1	1	227242.8	180089.0	1	227242.8
–	–	–	–	–	–	–	–	–	–	–
–	–	–	–	–	–	–	–	–	–	–

(continued)

Table 1.7 (continued)

8	1				**59924.9**	**0.99**	**98002.3**	130159.2	0.51	98002.3
	2				**60233.6**	**0.99**	**86573.0**	134789.5	0.99	86573.0
	3				**200869.0**	**1**	**567746.6**	272856.1	1	567746.6
	4				**92324.8**	**1**	**172701.7**	164700.4	1	172701.7
	5				**112905.1**	**1**	**227242.8**	186560.2	1	227242.8
–	–	–	–	–	–	–	–	–	–	–
–	–	–	–	–	–	–	–	–	–	–
12	1							**122853.6**	**0.99**	**98002.3**
	2							**127861.9**	**0.99**	**86573.0**
	3							**317014.0**	**1**	**567746.6**
	4							**166322.9**	**1**	**172701.7**
	5							**195481.8**	**1**	**227242.8**

Table 1.8 Daily Net Earnings (DNE) with Genco 3 learning

GenCo	Without Learning ($)	With GenCo 3 Learning ($)	With GenCo 3 Learning ($)	With GenCo 3 Learning ($)
GenCo 1	1556.47	0	0	0
GenCo 2	26.59	0	0	0
GenCo 3	56022.74	264412.73	607707.56	564220.68
GenCo 4	142.27	66962.03	250303.30	230070.08
GenCo 5	34267.60	33202.89	33097.68	33097.68
GenCo Cost Parameters				
GenCo 3 ai *($/MWh)*	25	50	91	92
GenCo 3 bi *($/MW²h)*	0.01	0.1	0.23	0.24

Table 1.9 Daily Net Earnings (DNE) with Genco 1 & 3 learning (Co-learning)

GenCo	Without Learning ($)	With GenCo 1 & 3 Learning ($)	With GenCo 1 & 3 Learning ($)
GenCo 1	1556.47	0	0
GenCo 2	26.59	0	0
GenCo 3	56022.74	607707.56	607707.56
GenCo 4	142.27	250303.30	250303.30
GenCo 5	34267.60	33097.68	33097.68
GenCo Cost Parameters			
GenCo 1 a_i ($/MWh)	*14*	*26*	*40*
GenCo 1 b_i ($/MW²h)	*0.005*	*0.13*	*0.2*
GenCo 3 a_i ($/MWh)	*25*	*91*	*91*
GenCo 3 b_i ($/MW²h)	*0.01*	*0.23*	*0.23*

other GenCos. The major objective of the GenCo is to achieve much improved net earnings by applying the co-learning effect using VRE-RL approach. Table 1.9 gives the daily net earnings (DNE) of GenCos with and without learning assuming GenCo 1 (Low cost, Low Capacity) and GenCo 3 (High Cost, High Capacity) having the learning capability.

The results indicate that GenCo 3 strategically uses its learning capability and increases their daily net earnings which is $607707.56. Even though GenCo 1 has the learning capability, it is not able to capitalize on it due to its wrong location of GenCo and its net earnings tend to zero as shown in Table 1.9 which in agent-based analysis is known as the substitution effect. Interestingly, the net earnings of non-learning GenCo 4 increase which in agent-based analysis is known as the complimentary effect. Later using the above way of approach, analysis is conducted assuming GenCo 3 (High Cost, High Capacity) and GenCo 5 (Low Cost, High Capacity) having the learning capability. Table 1.10 gives the daily net earnings (DNE) of GenCos 3 and 5 with and without learning.

The results indicate that both GenCo 3 and GenCo 5 are able to make use of its learning strategies and improve its daily net earnings. But the increase in net earnings of GenCo 5 is very less when compared to GenCo 5. Therefore, by applying multi-agent or co-learning approach, the learning capability applied to GenCo 3 and GenCo 5 made GenCo 1 and GenCo 2 to dispatch and increase its net earnings. An

Table 1.10 Daily Net Earnings (DNE) with Genco 3 & 5 learning (Co-learning)

GenCo	No Learning case ($)	With GenCo 3 & 5 Learning ($)	With GenCo 3 & 5 Learning ($)
GenCo 1	1556.47	5054.71	4640.73
GenCo 2	26.59	2408.40	2133.68
GenCo 3	56022.74	645984.99	672112.97
GenCo 4	142.27	305723.82	317662.63
GenCo 5	34267.60	38229.33	40414.11
GenCo Cost Parameters			
GenCo 3 a_i($/MWh)	*25*	*91*	*92*
GenCo 3 b_i($/MW²h)	*0.01*	*0.23*	*0.24*
GenCo 5 a_i ($/MWh)	*10*	*18.65*	*18.65*
GenCo 5 b_i($/MW²h)	*0.007*	*0.023*	*0.023*

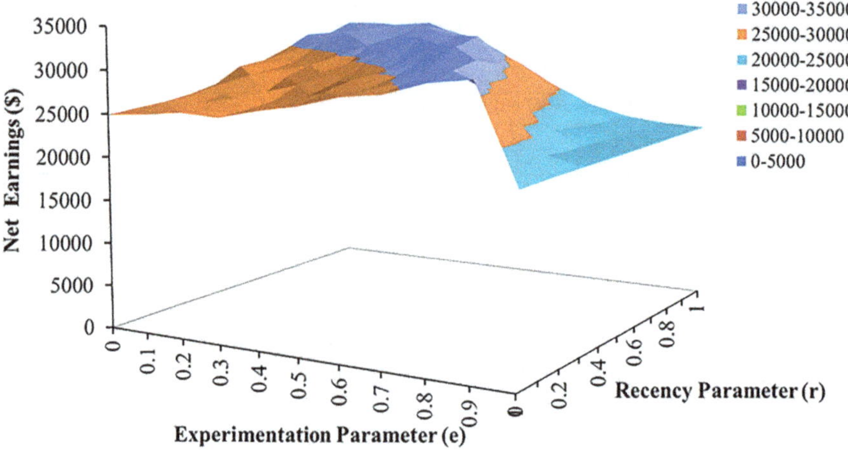

Fig. 1.15 Net earnings with high propensity

analysis is carried out on the GenCo 5 of the IEEE-5 bus system as a case study to make a choice on the better learning technique to check the variation of results obtained in both high propensity level and low propensity level using MRE-RL and VRE-RL technique.

For different experimentation (e) and recency (r) values, net earnings variation is taken. The net earnings variation with High Propensity and Low Propensity is shown in Figs. 1.15 and 1.16, respectively. Table 1.11 indicates the variation for both VRE and MRE.

It is clear that as degree of experimentation increases, there is considerable difference under high propensity case and maximum net earnings were obtained at $(e,r) = (0.7,0.4)$. For low propensity case, the maximum net earnings were obtained at $(e,r) = (0.9,1)$. Here high degree of experimentation and recency is required to

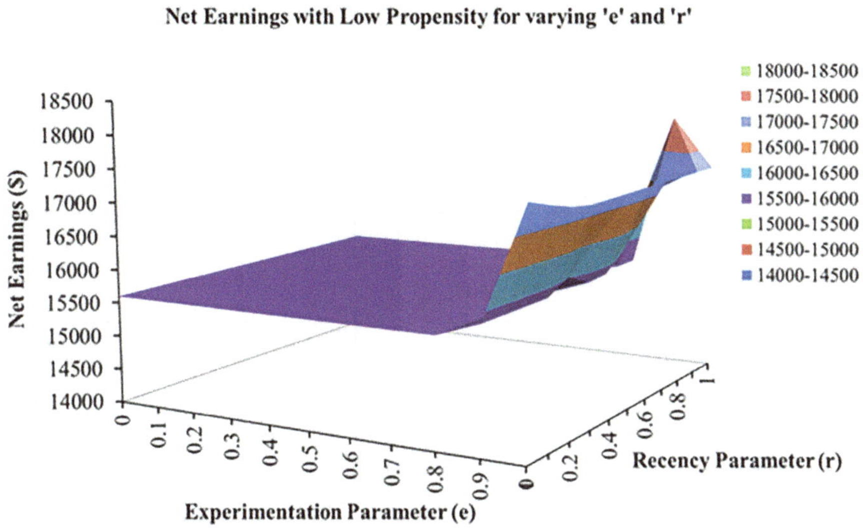

Net Earnings with Low Propensity for varying 'e' and 'r'

Fig. 1.16 Net earnings with low propensity

obtain considerably high net earnings. The analysis was later carried out for a period of 50 days and the GenCo commitments, Branch Power Flows and LMP for peak hour 17 are shown in Figs. 1.17, 1.18, and 1.19 respectively. The DNE of GenCo for a period of 50 days is shown in Fig. 1.20 which shows that GenCo 3 attains highest net earnings followed by GenCo 5.

Table 1.12 shows the reported cost parameters of all GenCos for selected Day 1, 10, 20, 30, 40 and 50. This clearly indicated the agent action taking place in the environment and the GenCos is updating cost parameters based on the previous day experience and its net earnings. From Fig. 1.17, it is clear that the commitment of GenCo 3 is high, since it is strategically learnt by reporting high cost parameters as evident from Table 1.12. This agent-based analysis can be extended to any number of days. Similarly, as specified by the user, the marginal pricing, commitments and net earnings can be found for any number of days. The stopping rule must be specified according to these criteria, and the entire operation is under the control of the independent authority. The action taken and the propensity will be different for all days. The probability of every action is according to the gain or loss of that particular agent on that day. The VRE interactive learning strategy applied in this agent-based analysis is a powerful learning algorithm which can adapt to any number of dynamically varying systems. The study clearly showed how GenCo commitments change with learning and corresponding effect in LMP values. Also, the VRE interactive learning algorithm helps GenCos to improve their net earnings and proves to be a good reinforced learning technique for a deregulated electricity market.

Table 1.11 Net earnings comparison with high initial propensity (VRE & MRE)

e→ r↓	0	0.1	0.2	0.3	0.4	0.5	0.6	0.7	0.8	0.9	1
0	24974	25352	25963	25795	27121	27946	28523	29414	30556	31091	20077
	24974	25353	25872	25670	26740	27595	28880	29447	30928	31590	21012
0.1	25198	25654	25800	26125	26794	28575	29390	30148	31620	24160	20805
	25198	25654	25798	25902	26674	28532	29191	30361	31704	29349	21084
0.2	25349	25494	26283	26507	27066	29701	30183	31351	32299	21781	20929
	25349	25529	26371	26375	27440	29310	29734	31461	32382	23283	21100
0.3	25771	26037	26355	28088	28162	29867	31406	33121	28025	21472	20981
	25771	26037	26157	27746	28348	29472	30940	33189	31551	21838	21098
0.4	26491	27989	27667	28809	29822	30990	33085	33166	23840	21359	21014
	26491	27894	27866	28774	29581	30561	32778	34017	26295	21583	21098
0.5	28244	28199	29051	29412	31471	33311	33830	29874	22532	21289	21017
	28244	28170	28931	28968	31090	33059	33560	31592	23247	21494	21096
0.6	28587	29325	30038	31721	33071	33001	32153	26748	22089	21296	21033
	28587	28832	29793	31147	32593	33322	32657	28752	22538	21416	21096
0.7	29710	30719	31493	31869	33522	32525	30255	24637	21890	21274	21044
	29710	30733	31730	31709	33504	32255	30658	25785	22195	21416	21095
0.8	31200	31763	31651	32944	33340	31492	27787	23717	21808	21222	21041
	31200	31703	31711	31999	32814	31651	28992	24377	21976	21346	21092
0.9	31018	31691	31971	33680	31574	29366	26040	23317	21687	21185	21038
	31018	31655	32633	33088	31229	29399	26688	23488	21914	21308	21092
1	32173	31840	32125	31094	29606	27411	24587	22922	21630	21185	21054
	32173	32302	31522	31616	29635	28009	25305	23082	21821	21276	21092

1.6 Conclusions

In this work, a new machine learning technique was applied to an electricity market for intelligent smart grid operation and analysed in detail to meet effective GenCo bidding strategy to maximize the profit. From GenCo point of view, the learning strategy is very much encouraging to maximize its net profit and helps towards smart scheduling. The cost parameter of GenCo without learning and the reported cost parameter through GenCo learning have improvised in achieving better net earnings. Through the interactive reinforcement learning approach, single ended bidding is efficiently handled to keep up the competitive interest of participation on day-ahead basis operation in wholesale electricity market. This type of agent-based modelling of electricity systems is quite suitable for wholesale power market with load serving entities comprising of industrial, commercial and domestic consumers. Also, the market entities that are responsible for creating the line congestion only

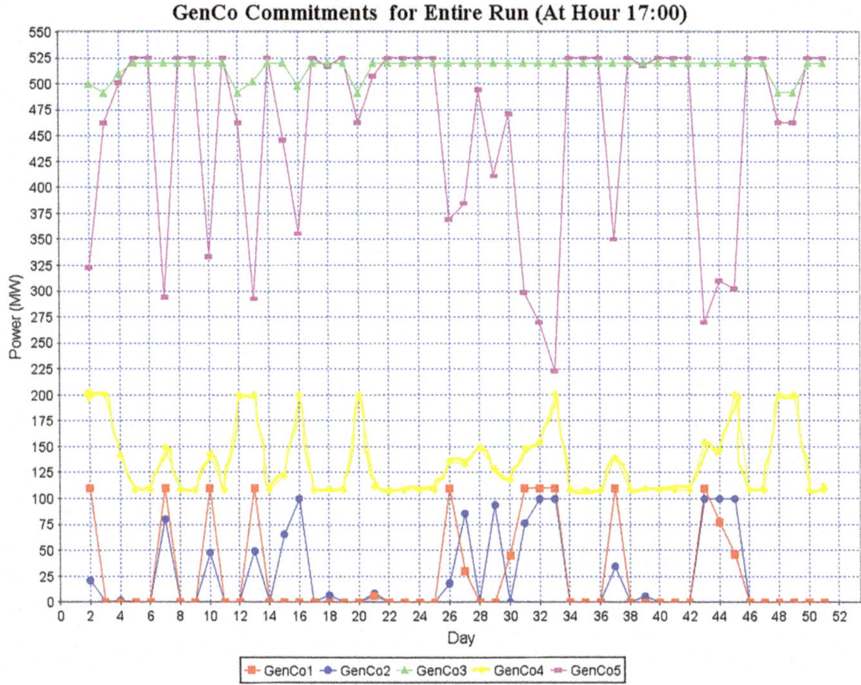

Fig. 1.17 GenCo commitments (50 Days)

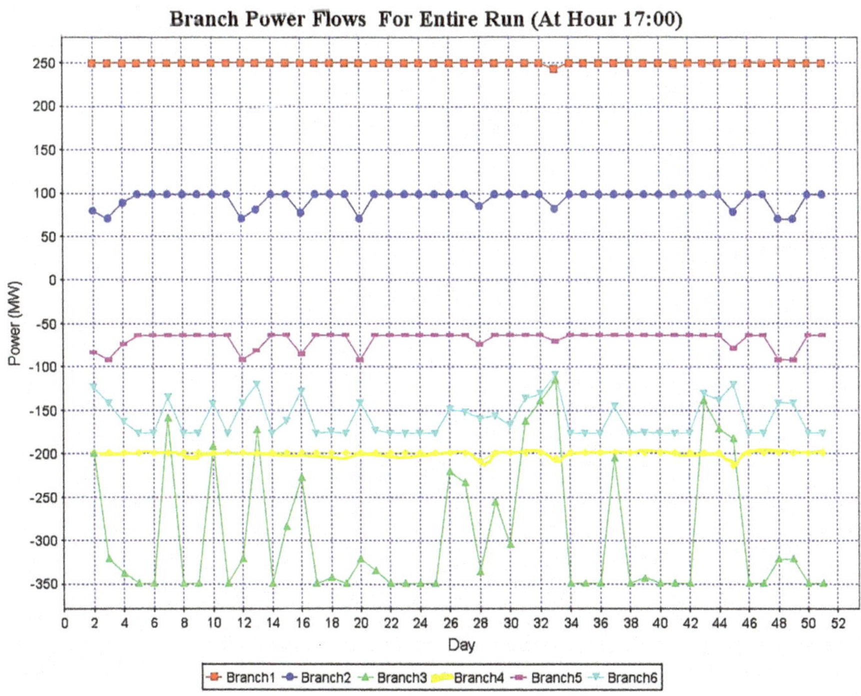

Fig. 1.18 Branch power flow (50 Days)

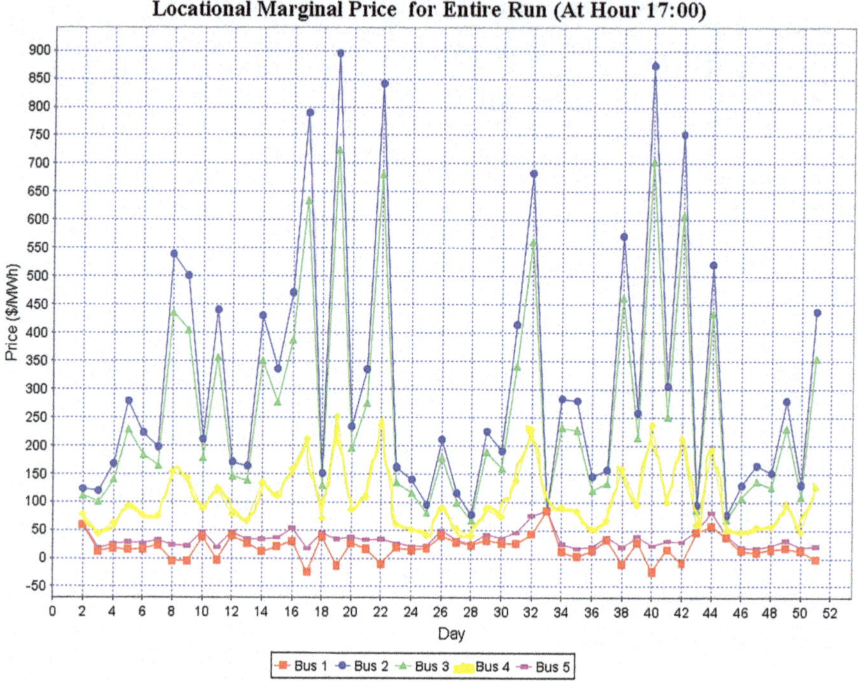

Fig. 1.19 LMP variation (50 Days)

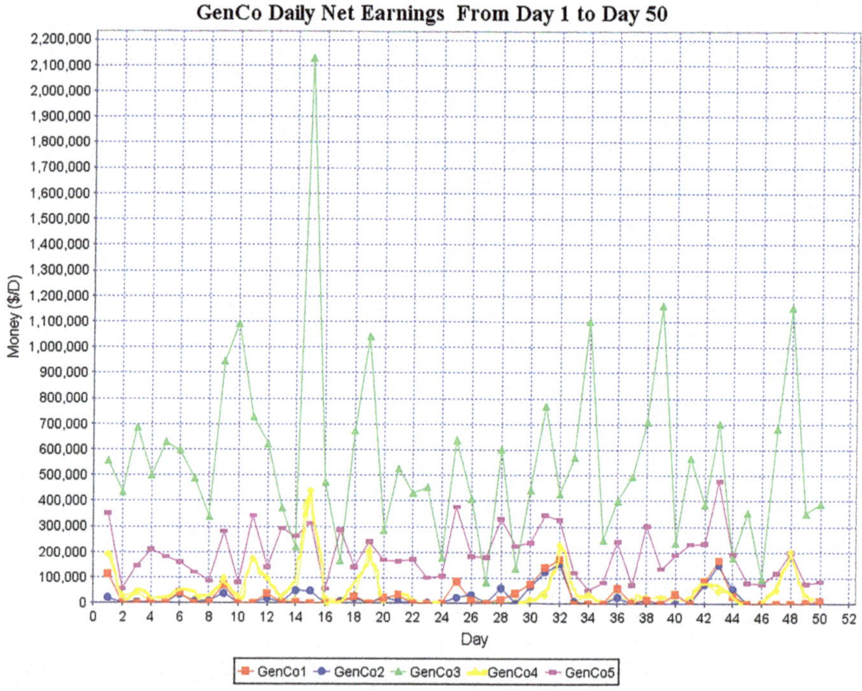

Fig. 1.20 DNE of GenCos (50 Days)

Table 1.12 Reported cost parameters for interactive VRE learning

GenCo	Day	ai^R (\$/MWh)	bi^R (\$/MW^{2}h)
GenCo 1	1	15.2727	0.0496
	10	56	0.2545
	20	15.2727	0.0972
	30	14	0.0279
	40	18.6667	0
	50	15.2727	0.0139
GenCo 2	1	36	0.54
	10	18	0.03
	20	15	0.087
	30	16.3636	0.0584
	40	30	0.3
	50	20	0
GenCo 3	1	75	0.0361
	10	100	0.0481
	20	33.3333	0.0133
	30	30	0.0165
	40	37.5	0.0072
	50	33.3333	0.0263
GenCo 4	1	40	0.0333
	10	60	0.3
	20	51.4286	0.2571
	30	120	0.06
	40	72	0.1286
	50	120	0.0273
GenCo 5	1	40	0.0333
	10	15	0.0042
	20	10.9091	0.0216
	30	17.1429	0.047
	40	24	0.0067
	50	17.1429	0.0041

need to pay the congestion price which is helpful for the load serving entities to get the power at low rate. In this work, an IEEE 5-bus test system is considered for the entire analysis and a detailed investigation has been done taking into account the GenCo commitment, LMP, GenCo Net earnings, Co-learning effects, Microgrid integration and extension to 50 days learning. The application of VRE learning strategy in a GenCo is also detailed for high and low propensity cases with varying experimentation and recency parameters. When compared to MRE and RE, the probability calculation which fails to converge, VRE algorithm allows accurate probability calculation through interactive approach on GenCos even if the difference between the propensities is less to perform smart bidding and intelligently schedule the GenCos to achieve maximum profit. These have been carried out to prove the effectiveness of agent-based computational economics-based method. The agents with its intelligence and interactive ability help the deregulated electricity market to evolve for taking suitable actions under single ended bidding environment. As a future work, extended analysis can be carried out on double ended bidding process as well as handling the retail electricity market. The level of clearing price and market power can be analysed and compared under varying market mechanisms which may be suitable for wholesale electricity market scenario. Further, even the learning strategy can be modified utilizing algorithm like deep evolutionary reinforced learning on the realistic grid network. The platform used in this work can be integrated with GridLab-D so as to incorporate more distributed generation and data-driven method like Deep Reinforcement Learning techniques. The applied model incorporates learning capability so that the GenCo agent is able to achieve high net earnings through smart bidding which is advantageous to the utility companies. A similar approach can be applied to TransCos too, but it will be a challenging task in handling dynamically changing power grid.

References

1. M. Shahidehpour, H. Yamin, Z. Li, *Market Operations in Electric Power Systems: Forecasting, Scheduling and Risk Management* (Wiley-IEEE Press, 2002)
2. J. Antonio, Conejo, electricity markets: Analysis & operations. IET Gener. Transm. Distrib. **4**, 123–124 (2010)
3. H. Liu, L. Tesfatsion, A.A. Chowdhury, Derivation of locational marginal prices for restructured wholesale power markets. J. Energy Mark **2**(1), 3–27 (2009)
4. L. Deng, Z. Li, H. Sun, Q. Guo, X. Yinliang, R. Chen, J. Wang, Y. Guo, Generalized locational marginal pricing in a heat-and-electricity-integrated market. IEEE Trans. Smart Grid **10**(6), 6414–6425 (2019)
5. A. Radovanovic, T. Nesti, B. Chen, A holistic approach to forecasting wholesale energy market prices. IEEE Trans. Power Syst. **34**(6), 4317–4328 (2019)
6. P. Andrianesis, M. Caramanis, R.D. Masiello, R.D. Tabors, S. Bahramirad, Locational marginal value of distributed energy resources as non-wires alternatives. IEEE Trans. Smart Grid **11**(1), 270–280 (2020)
7. S. Huang, W. Qiuwei, H. Zhao, C. Li, Distributed optimization based dynamic tariff for congestion management in distribution networks. IEEE Trans. Smart Grid **1** (2017)
8. Mehdi Rahmani- Andebili, Optimal operation of a plug-in electric vehicle parking lot in the energy market considering the technical, social, and geographical aspects, in *Planning and Operation of Plug-in Electric Vehicles*, (Springer, Cham, 2019), pp. 105–147
9. M. Rahmani-Andebili, G.K. Venayagamoorthy, Investigating effects of changes in power market regulations on demand-side resources aggregators, in IEEE Power & Energy Society General Meeting, Denver, pp. 1–5, 20–30 July 2015
10. Z. Zhou, A. Botterud, Dynamic scheduling of operating reserves in co-optimized electricity markets with wind power. IEEE Trans. Power Syst. **29**(1), 160–171 (2014)
11. S.D.J. McArthur, E.M. Davidson, V.M. Catterson, A.L. Dimeas, N.D. Hatziargyriou, F. Ponci, T. Funabashi, Multi-agent systems for power engineering applications-part I: Concepts, approaches and technical challenges. IEEE Trans. Power Syst. **22**(4), 1743–1752 (2007)
12. P. Kiran, K.R.M. Vijaya Chandrakala, T.N.P. Nambiar, Multi-agent based systems on micro grid—A review, in *IEEE International Conference on Intelligent Computing and Control*, (2017)

13. G. Conzelmann, G. Boyd, V. Koritarov, T. Veselka, Multiagent power market simulation using EMCAS, in *IEEE Power Engineering Society Gen. Meeting*, (2005), pp. 2829–2834
14. T. Sueyoshi, G.R. Tadiparthi, A wholesale power trading simulator with learning capabilities. IEEE Trans. Power Syst. **20**(3), 1330–1340 (2005)
15. Z. Vale, T. Pinto, I. Praca, H. Morais, MASCEM: Electricity markets simulation with strategic agents. IEEE Trans. Intell. Syst. **26**(2), 9–17 (2011)
16. W. Ketter, J. Collins, P. Reddy, Power TAC: A competitive economic simulation of the smart-grid. Energy Econ. **39**, 262–270 (2013)
17. D. Krishnamurthy, W. Li, L. Tesfatsion, An 8-zone test system based on ISO New England data: Development and application. IEEE Trans. Power Syst. **31**(1), 234–246 (2016)
18. M. Kaveh Dehghanpour, H. Nehrir, J.W. Sheppard, N.C. Kelly, Agent-based modeling of retail electrical energy markets with demand response. IEEE Trans. Smart Grid **9**(4), 1–11 (2018)
19. P. Kiran, K.R.M. Vijaya Chandrakala, T.N.P. Nambiar, Agent based locational marginal pricing and its impact on market clearing price in a deregulated electricity market. J. Electr. Syst. **15**(3), 405–416 (2019)
20. P. Kiran, K.R.M. Vijaya Chandrakala, Roth-Erev reinforcement learning approach for smart generator bidding towards long term electricity market operation using agent based dynamic modeling. Electr. Power Compon. Syst., Taylor & Francis **48**(3), 256–267 (2020)
21. P. Kiran, K.R.M. Vijaya Chandrakala, New interactive agent based reinforcement learning approach towards smart generator bidding in electricity market with micro grid integration. Appl. Soft Comput. J., Elsevier **97**(Part A), 106762 (2020)
22. J. Yang, J. Zhao, F. Luo, F. Wen, Z.Y. Dong, Decision-making for electricity retailers: A brief survey. IEEE Trans. Smart Grid **9**(5), 4140–4153 (2017)

Chapter 2
Reinforcement Learning Techniques for MPPT Control of PV System Under Climatic Changes

Maximiliano Trimboli, Luis Avila, and Mehdi Rahmani-Andebili

Abstract Photovoltaic (PV) systems have become a potential solution to global problems like pollution and climate changes resulting from the excessive use of fossil fuels. This kind of system can respond to the constant increase in the electric energy demand and the need for energy supply in rural or hard-to-reach areas. However, as the energy efficiency of PV systems is low, there exists a necessity to maximize the output power so that it reaches the maximum power point (MPP). Different Maximum Power Point Tracking (MPPT) techniques can be used to increase the efficiency of PV systems. Nevertheless, climatic variations make their task difficult to achieve. This work proposes the use of reinforcement learning (RL) techniques for solving the MPPT problem of a PV system under different conditions of temperature and solar irradiance. RL techniques do not require information of a model that describes the behavior of the system with its environment. They only make use of the information of the possible states to visit and actions to take and updates a utility function according to how good the last action taken was. To validate the effectiveness of the proposed algorithms, several experiments were performed in a simulated environment. The obtained results show good performances with stable behaviors, proving to be practical for the control of photovoltaic systems.

Keywords Photovoltaic Systems · MPPT · Reinfocement Learning · Variability

M. Trimboli · L. Avila (✉)
Laboratorio de Investigación y Desarrollo en Inteligencia Computacional (LIDIC), CONICET-UNSL, San Luis, Argentina
e-mail: mdtrimboli@unsl.edu.ar; loavila@unsl.edu.ar

M. Rahmani-Andebili
Electrical Engineering Department, Montana Technological University, Butte, MT, USA
e-mail: mrahmaniandebili@mtech.edu

M. Rahmani-Andebili (ed.), *Applications of Artificial Intelligence in Planning and Operation of Smart Grids*, Power Systems,
https://doi.org/10.1007/978-3-030-94522-0_2

2.1 Introduction

Solar energy is the most used renewable energy source around the world, and it is technically inexhaustible, secure, easily accessible, not noisy, or polluting [1]. It can be exploited using photovoltaic (PV) systems through the photoelectric effect, converting absorbed solar radiation into electrical energy. This method is known as direct solar energy conversion [2]. Photovoltaic systems have become a potential solution to global problems like climate changes and pollution, and lack of supply from electric grid in rural zones or areas with difficult access due to the high cost of installation. In particular, PV systems are a promising alternative for electric energy generation as they are systems characterized by being reliable, low cost, and easily implementable.

However, they turn into unstable systems in relation to their efficiency, due to high dependence on uncertain climatic conditions [3, 4]. On the other hand, renewable energy sources can be integrated with energy storage systems to mitigate their power variability and make them controllable and dispatchable [5, 6]. In this sense, to be efficient, a PV panel must transmit continuously the maximum power available to the load under variable conditions, so this control problem is known as "Maximum Power Point Tracking" (MPPT). When the system operates near its maximum efficiency, the total system costs reduce significantly.

In order to optimize the performance of photovoltaic applications, numerous MPPT control approaches have been proposed in the scientific literature. On the one hand, indirect approaches are based on data from precalculated Power-Voltage (P-V) curves for different environmental conditions or on experimentally obtained mathematical models. Among the most widely used indirect methods are those based on the open circuit voltage (Voc) and the short-circuit current (I_{SC}), where the calculations to obtain the maximum power point (MPP) are based on the values of these variables [7, 8]. Similarly, lookup table-based methods compare voltage and current values measured with stored MPPs for given environmental conditions, making use of numerical models to approximate the behavior of the PV source [9, 10]. The main advantage of these methods is their simplicity; however, they cannot be easily adapted to any external changes to the PV source.

On the other hand, direct methods are based on current and voltage measurements and have the advantage of being independent of the PV source used. Among the most popular methods in this category, we can mention perturb and observe (P&O) methods [11, 12]; incremental conductance (IC) methods [13, 14]; those based on fuzzy logic [15, 16]; and neural networks [17, 18]. The P&O and IC methods have the advantage of having low computational complexity at the cost of fluctuations around the MPP during steady state and lack of robustness in the face of changes in environmental conditions. Even though fuzzy logic and neural networks techniques are more robust, they depend on a priori knowledge, which increases the complexity of the implementation. Fuzzy logic is a technique that does not need precise mathematical model and is normally used in cases where the process is binary form and is not applicable [16, 19]. In turn, neural networks are methods of

Machine Learning that simulates the learning mechanism of biologic neurons for the search of solutions to nonlinear systems. They consist of a layer that receives inputs, a hidden layer that obtains the data from the input layer, and sends it to an output layer that generates a response to the system [20, 21].

Recently, some works have made use of reinforcement learning (RL) techniques seeking to mitigate the problems mentioned in the MPPT control [1, 22]. Reinforcement learning brings together numerous algorithms that learn through interaction with the environment how to achieve a complex goal or maximize the accumulated value of the rewards obtained along a particular dimension [23].

Here, we present the use of computational intelligence techniques to improve the performance of PV systems. More precisely, we focus on model-free approaches, that is, those that do not require model knowledge or system parameters. RL techniques build a control policy from trajectories obtained from interactions with the environment while searching for maximizing the accumulative value of futures utilities. In particular, reinforcement learning (RL) techniques are a promising alternative for MPPT control, regardless of the characteristics of PV systems and variable climatic conditions.

The work is organized as follows. In Sect. 2.2, we introduce the PV system modeling based on the equivalent circuit method. Section 2.3 presents reinforcement learning techniques and their theoretical bases. In Section 2.4, several experiments are implemented on the PV simulated model, and results are analyzed. Finally, Section 2.5 presents some conclusions and final remarks.

2.2 PV Modeling

2.2.1 Introduction to PV Systems

A PV system is a set of cells whose aim is to generate a desired electrical power from the absorption of solar radiation through the photoelectric effect. The fundamental element of a PV system is a PV cell, which consists of thin layers of semiconductor material (generally silicon) that define a p-n union. A cell absorbs an incident flow of photons, generating electron-hole pairs separated by the electric field at the junction of the semiconductor, producing an electric current. This current is proportional to the incident solar radiation and flows through the load of the external circuit [24].

A solar cell can generate between 1 and 2 Watts depending on the semiconductor material. Therefore, to obtain higher output power from the PV system, many cells are interconnected in series to form a PV module. Also, modules can be interconnected in parallel to form a PV matrix or array. According to Fig. 2.1, this is because the number of connected elements (cells or modules) in series is directly proportional to the output voltage of the PV system, and the number of elements in parallel is directly proportional to the output current [25].

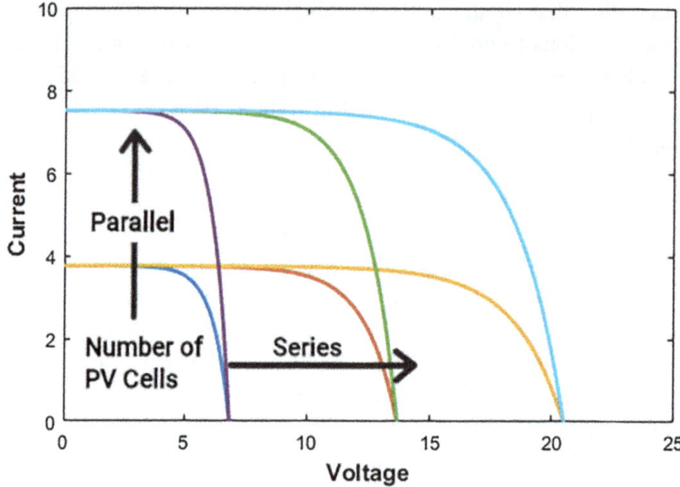

Fig. 2.1 Behavior of current-voltage curves in relation to the number of PV cells in a PV system

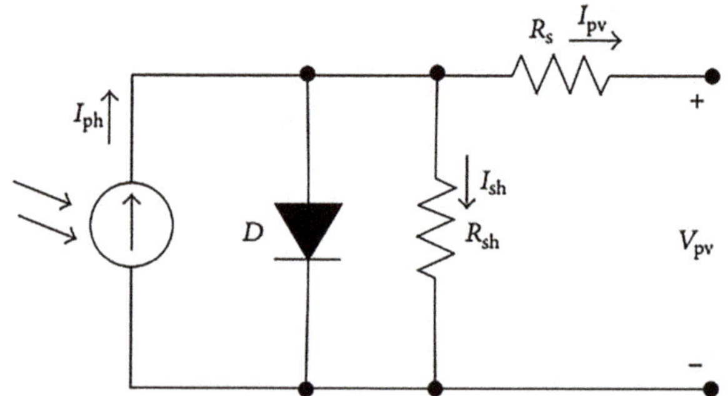

Fig. 2.2 Equivalent circuit of one-diode model

2.2.2 *PV Cell Model*

The PV cell model is the most important aspect to consider for simulating the behavior of PV modules or arrays, as depending on the selection of pre-established models, the precision of the entire system varies [26]. The general model of a PV cell, also known as the one-diode-model, consists of a current source, a diode, a parallel resistance (shunt resistance) that represents the current losses, and a series resistance that represents the opposition to the current flow, as seen in Fig. 2.2 [27].

According to Kirchhoff's current laws, the output current of the equivalent circuit in Fig. 2.2. can be expressed as follows [28]:

$$I_{pv} = I_{ph} - I_D - I_{sh} = I_{pv} - I_D - \frac{V_{pv} + I_{pv} R_s}{R_{sh}}, \tag{2.1}$$

where I_{ph} is the photo-generated current or photocurrent, I_D is the current that circulates through the diode D, and I_{sh} is the current that passes through the shunt resistance, which depends on the output voltage V_{pv}, the output current I_{pv}, the series resistance R_s, and the shunt resistance R_{sh}, as shown in Eq. 2.1. In the simplest one-diode model, losses are not considered; thus, R_{sh} is infinite and R_s is equal to zero [1, 27, 29].

Here, we use a model where only R_{sh} is infinite, resulting in the circuit shown in Fig. 2.3. Therefore, I_{pv} can be expressed as the difference between I_{ph} and I_D.

By Shockley's diode equation, the current I_D can be expressed as follows:

$$I_D = I_o \left(e^{\frac{qV_D}{\eta KT}} - 1 \right) = I_o \left(e^{\frac{q(V_{pv} + I_{pv} R_s)}{\eta KT}} - 1 \right), \tag{2.2}$$

where I_o is the reverse saturation current, q represents the electron charge in Coulomb, V_D is the diode voltage, η represents the diode ideality factor (For an ideal diode, $\eta = 1$), K is Boltzmann constant, and T represents the cell temperature expressed in Kelvin [29, 30]. The quotient $\eta KT/q$ is known as thermal voltage of the diode.

Replacing (2.2) into equation based on Kirchhoff's current laws for Fig. 2.3, the characteristic equation of the output current is obtained for the model used:

$$I_{pv} = I_{ph} - I_o \left(e^{\frac{q(V_{pv} + I_{pv} R_s)}{\eta KT}} - 1 \right) \tag{2.3}$$

The output voltage and current are given by the following equations:

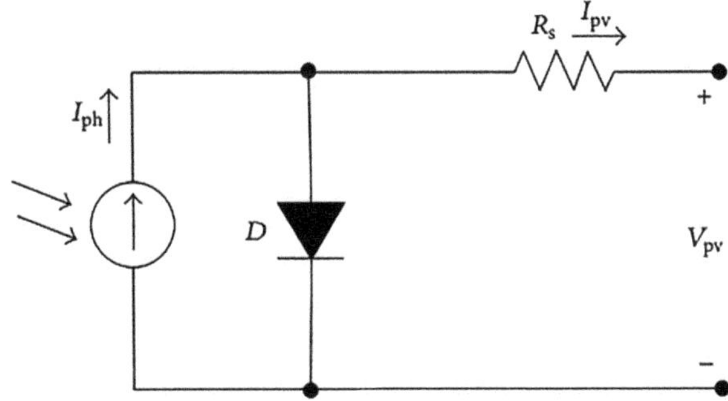

Fig. 2.3 Equivalent circuit of one-diode model without the shunt resistance

$$V_{pv} = \frac{q}{\eta KT} \ln\left(\frac{I_{ph} - I_{pv} - I_o}{I_o}\right) - I_{pv} R_s \tag{2.4}$$

$$P_{pv} = I_{pv} V_{pv} = I_{pv} \left[\frac{q}{\eta KT} \ln\left(\frac{I_{ph} - I_{pv} - I_o}{I_o}\right) - I_{pv} R_s\right] \tag{2.5}$$

In this way, the photocurrent I_{ph} is obtained by the following equation:

$$I_{ph} = I_{ph(T_{ref})} \left(1 + K_0 \left(T - T_{ref}\right)\right), \tag{2.6}$$

where K_0 represents the temperature coefficient for the short-circuit current and T_{ref} is the reference temperature of the cell. In turn, the photocurrent at reference temperature is described as follows:

$$I_{ph(T_{ref})} = \frac{G}{G_{(nom)}} I_{SC(Tref,nom)} \tag{2.7}$$

We can observe in (2.6) and (2.7) that I_{ph} depends on the irradiance G, which can be defined as the amount of light energy from an object hitting a square meter of another, each second [31], expressed in W/m^2, and the cell temperature. The reverse saturation current can be obtained by the following:

$$I_o = I_{o(Tref)} \left(T / T_{ref}\right)^{3/n} * e^{-\frac{qV_g}{\eta k}\left(\frac{1}{T} - \frac{1}{T_{ref}}\right)} \tag{2.8}$$

$$I_{o(Tref)} = \frac{I_{SC(Tref)}}{\left(e^{-\frac{qV_{oc(Tref)}}{\eta k T_{ref}}} - 1\right)}, \tag{2.9}$$

where $I_{SC(Tref)}$ is the short-circuit current of the PV cell at reference temperature and radiation, V_G represents the energy of the prohibited band of the semiconductor [32]. Finally, $V_{OC(Tref)}$ is the open-circuit voltage at reference temperature.

2.2.3 PV System Model

In Fig. 2.4, the equivalent circuit for a PV system is shown. As mentioned in Sect. 2.2.1, PV systems are composed of a certain number of cells connected in series and parallel. In the circuit, those cells are represented by the same number of diodes.

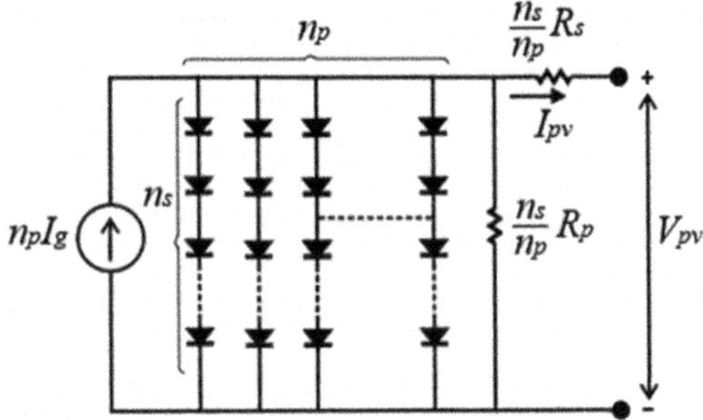

Fig. 2.4 Equivalent circuit of PV system

Thus, N_p defines the number of diodes (or cells) connected in parallel and N_s represents the number of diodes connected in series.

The characteristic equation of a photovoltaic system is given as follows:

$$I_{pv} = N_p I_{ph} - N_p I_o \left(e^{\frac{1}{V_t}\left(\frac{V_A}{N_s}+\frac{I_A}{N_p}R_s\right)} - 1 \right) - \frac{N_p}{R_{SH}}\left(\frac{V_A}{N_s}+\frac{I_A}{N_p}R_s\right) \tag{2.10}$$

The total output of PV system, also known as operation point, is expressed as follows:

$$P_{pv} = V_{pv} I_{pv} \tag{2.11}$$

The equivalent electric circuit shown in Fig. 2.4. and its characteristic Eq. (2.10) can be manipulated to reduce the system to a module or a cell, or it can be considered as an array. To consider a PV cell, we assume $N_p = N_s = 1$, $N_p = 1$ for a PV module, and N_p as N_s must be >1 for an array.

From the above equations, we can obtain the current-voltage (I-V) and the power-voltage (P-V) curves, which are called characteristic curves, because they describe a PV system. Considering the case of a single module, we will obtain characteristic curves like those shown in Fig. 2.5. The I-V (blue) curve corresponds to current values for each voltage value between 0 and 25 Volts, while the P-V (red) curve represents power values for the same range.

2.2.4 *Maximum Power Point*

The Maximum Power Point (MPP) of a PV system is a unique point where output power is maximized, given the product between the V_{MPP} and I_{MPP}, which depends on the temperature and the solar radiation absorbed [25, 32]. To increase efficiency, the operating point must approach the MPP, mathematically that means that the condition

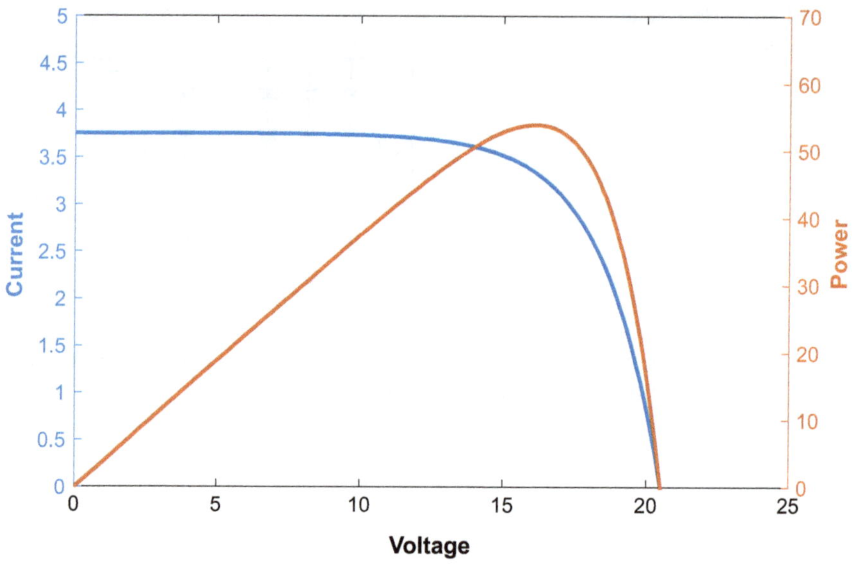

Fig. 2.5 I-V and P-V characteristic curves of PV module

$$\frac{dP_{PV}}{dV_{PV}} = 0 \tag{2.12}$$

must be fulfilled [33].

When a load is connected to a PV system, the operating point is defined by the intersection of the load curve with the I-V curve of the PV system [34]. Consequently, as the load varies, the operating point also does. However, the MPP does not always remain stable at the same value due to climatic variations that disturb the point [25, 30]. Figures 2.6 and 2.7 show the typical curves of the model, the irradiance level is constant, while the temperature varies and then the temperature is constant, while the irradiance varies, respectively. As can be seen, when the environment conditions change, the curves, and therefore the MPP's, change. An increase in the ambient temperature causes a drop in the maximum power available, while an increase in the irradiance generates a rise in the maximum power available.

For the reasons mentioned above, it is necessary to monitor the MPP using MPPT techniques and take control actions on the operating point [23]. There are different types of MPPT techniques in the literature and each of them has greater efficiency at the cost of greater complexity.

Fixed Duty Cycle is considered as the simplest MPPT method, since it only consists of adjusting the load impedance to achieve the MPP [35, 36]. Beta methods make an approximation based on (2.13) applied on a conventional closed circuit with constant reference.

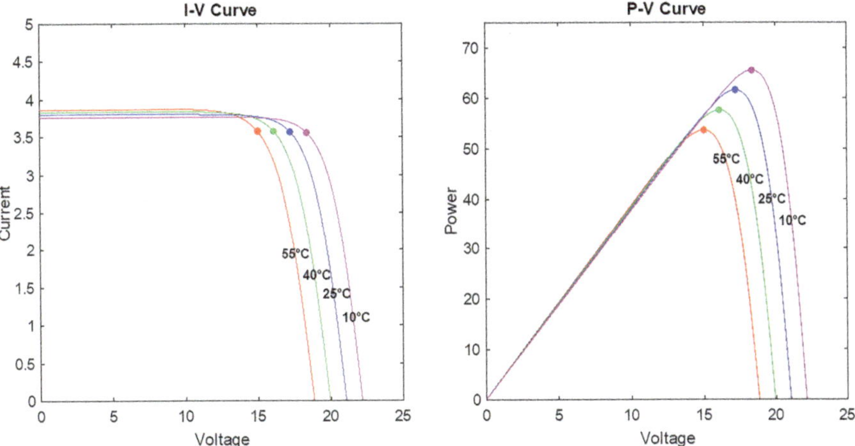

Fig. 2.6 I-V and P-V curves at constant irradiance

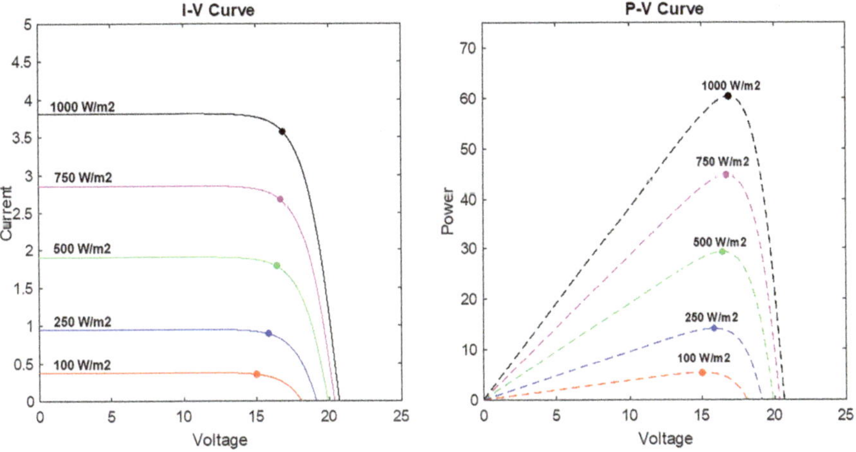

Fig. 2.7 I-V and P-V curves at constant temperature

$$\beta = \ln\left(\frac{I_{\mathrm{PV}}}{V_{\mathrm{PV}}}\right) - \frac{q}{\eta KTN_{\mathrm{s}}} V_{\mathrm{PV}} \tag{2.13}$$

P&O methods, also known as *"Hill Climbing,"* are the most used for their ease of implementation. The output power is periodically tested using the output voltage and current values of the module and compared with the power in the previous time step, increasing or decreasing the input voltage to approach the MPP [37]. If the perturbation leads to an increase or decrease in the panel power, the subsequent perturbation is made in the same or the opposite direction, respectively. In this manner, the peak power tracker continuously seeks the peak power condition.

IncCond is more complex than P&O: it is based on the derivative of the curve to determine the position of the operating voltage with respect to MPP, according to (2.14). For this reason, it requires not only the measured voltage and power but also the current. If the difference between voltage and current measured in a time interval is equal to zero, it means that the operating point is stabilized on the MPP; otherwise, depending on its values and the slope $\Delta P/\Delta V$, a positive or negative disturbance is executed on the operating voltage of the module [37, 38].

$$\frac{dI}{dV} = -\frac{I}{V}; \frac{dP}{dV} = 0, \text{at MPP} \tag{2.14a}$$

$$\frac{dI}{dV} > -\frac{I}{V}; \frac{dP}{dV} > 0, \text{left of MPP} \tag{2.14b}$$

$$\frac{dI}{dV} < -\frac{I}{V}; \frac{dP}{dV} < 0, \text{right of MPP} \tag{2.14c}$$

Constant voltage and current are techniques that assume linear dependence between V_{MPP} or I_{MPP}, and the open-circuit voltage. To reach the real maximum power, environmental conditions must be kept constant; otherwise, it is necessary a periodic mechanism to search V_{oc} [7, 21]. If the voltage is constant, that dependence is expressed as follows:

$$\frac{V_{\text{MPP}}}{V_{\text{OC}}} \cong K < 1 \tag{2.15}$$

2.3 MPPT with Reinforcement Learning

2.3.1 Introduction to Reinforcement Learning

Reinforcement learning is a computational intelligence method in which initially the learner –located in an environment – does not know what actions to take to maximize a value function. Therefore, the agent must discover through trial and error which actions maximize the benefits, considering that the highest reward is obtained after a succession of actions. In RL, there exists an interaction with a real or simulated uncertain environment. Hence, an RL algorithm results useful when there is not a model of the environment and when no rules exist to choose the most appropriate action in each environment state [23, 40]. For this reason, RL has been successful in a variety of applications where the environment is modified over time, such as chess [41], video games [42], mobile robotics [43], autonomous vehicles [44], control of active and dynamic variables [45], etc.

An RL problem describes an agent that decides what actions to take based on the information he or she perceives from the space it interacts with. Within this problem, there are four elements to consider:

- A rule is chosen by the agent to select a certain action in each state. It is called police and is considered as a mapping between the actual state and possible actions to be chosen. The objective of RL is to reach that the chosen police generates a maximization of accumulated reward.
- A reward signal that quantifies how good was the last action executed by the agent. The value received for each transition from one state to another is called reward and can be positive or negative (penalization) depending on if the agent follows a correct behavior or not.
- A value function that determines how good is in the long term, a trajectory to take from a certain state.
- For RL problems where model-based methods are used, there is a model that defines the behavior of the environment or infers how it will behave.

As already mentioned above, the agent tries to maximize reward obtained over time by executing actions that it knows from experience that will get it to that objective. This process is known as exploitation. That experience has been acquired in the middle of the learning method through a process called exploration, which consists of test actions that have not been previously chosen under the state in which the agent is at that moment and that does not necessarily seek to exploit the knowledge already learned. In RL, it is not possible to work exclusively with one of these ways of operating without failing to solve the problem; thus, we need a balance between exploration and exploitation. A pure exploration will not allow learning the optimal route to reach the objective, while pure exploitation will not access new and better knowledge.

There are algorithms dedicated to finding a solution to this dilemma, which are incorporated into the RL, for example, epsilon greedy or ε-greedy policy, that defines a probability value of inclination to exploitation ($0 \leq \varepsilon \leq 1$). Thus, the algorithm will choose to explore or exploit depending on the previous value obtained to the decision of certain action. To capture the most important aspects of the real task, it is necessary to formalize using a theoretical dynamic system through Markov decision processes.

2.3.2 *Markov Decision Processes*

To formalize an RL problem through Markov decision processes (MDP) [23], it is necessary to satisfy the Markov property. The Markov property establishes that the action executed by the agent does not depend only on the actual state of the environment, which means that it does not matter what actions were taken previously and information contained in the actual state is sufficient to decide what future action to take. If the number of states and actions in an RL problem is finite, it is said that the MDP model is finite as well [46].

2.3.2.1 MDP Formulation

Generally, in RL formulations with MDP, the problem is defined by four elements, which form a tuple (*S*, *A*, *p*, *R*) [23, 40], where

- *S* is a set of states, that is, a set of situations that are derived from an action or previous actions. Each state s_i is an element that belongs to *S*:

$$S = \left[s_0, s_1, s_2, \ldots, s_n\right] \subseteq \mathbf{s} \tag{2.16}$$

- *A* is the set of actions that the agent uses to generate a state transition based on a certain policy:

$$A = \left[a_0, a_1, a_2, \ldots, a_n\right] \subseteq \mathbf{a} \tag{2.17}$$

- The reward signal *R* quantifies the preferences of the agent giving a numeric value when changing state due to the action executed by the agent. Therefore, the reward signal is expressed as shown in (2.18).

$$R : S \times A \rightarrow r \tag{2.18}$$

- A state transition function *p* that determines the probabilities of going from one state to each of the possible states, and defines the MDP dynamic. (2.19) defines probability of proceeding from state *S* to state *S′* executing action *A*.

$$p\left(s', a, s\right) = p\left(s' | , s | , a\right) \tag{2.19}$$

Figure 2.8. shows the agent-environment, where at time step *t*, the agent takes an action a_t ϵ *A* based on the perception of the previously received information from the environment about the state s_t ϵ *S* where the agent is located and the reward r_t ϵ *R*. Then, at the next time step *t* + *1*, the environment gives the new reward (r_{t+1}) and state (s_{t+1}), according to the corresponding action taken. Through time *n*, the following sequence or trajectory is formed: s_0, a_0, r_1, s_1, a_1, r_2, a_2, s_2…r_n, s_n, a_n.

For efficient decision-making, it is necessary to find a policy *π* that will maximize the function used to the accumulated rewards [47]. It can be deterministic, that

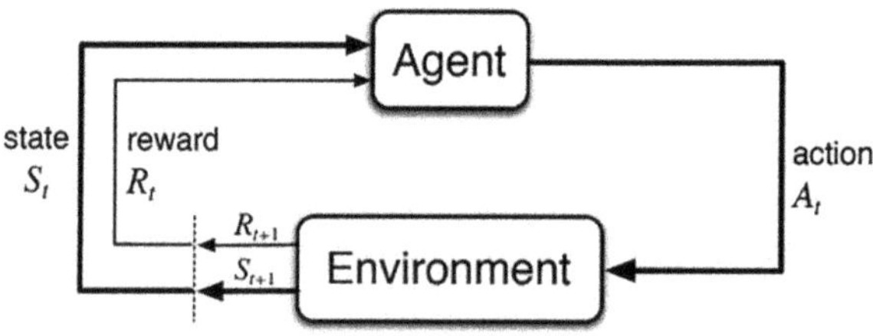

Fig. 2.8 Markov agent–environment interface

is, the agent will always choose the same action in a particular state; or it can be stochastic, where the policy defines the probability of taking one action or another, in the state *st* and time *t*. Hence, in a stochastic policy, the agent's actions may vary under the same conditions. Besides, the policy can be stationary (variable in time) or not stationary (constant in time).

For random variables that have a discrete probability distribution, the next state only depends on the previous state and action. This means that, for the values given above, the probability that these values occur in time t is given by the transition function p and is expressed in the following equation [17].

$$p(s',r|s,a) \doteq \Pr\{s_{t+1} = s', r_{t+1} = r \mid s_t = s, a_t = a\} \tag{2.20}$$

2.3.2.2 Episodes and Returns

Before the execution of an action, the agent should not only take into account the next reward but also the future expected rewards that it will receive from the environment following the path given by that action. For example, in the first time steps, the rewards obtained are insignificant, but the agent, through its experience, knows that the sum of the rewards corresponding to the path taken will be much greater than other paths where the instant reward is greater. That accumulated reward is called the return and it is expressed as (2.21), where T is the last time step [40].

$$G_t \doteq R_{t+1} + R_{t+2} + R_{t+3} + \ldots + R_T = \sum_{k=0}^{T} R_{t+k+1} \tag{2.21}$$

As reflected in (2.21), all rewards that form the return keep their value, regardless of the time the agent is. However, for cases where the time step number is too high, the return may grow infinitely. To avoid this and consider the importance of the time in which these rewards were obtained, the variable knows as the expected return is calculated as (2.22).

$$G_t \doteq R_{t+1} + \gamma R_{t+2} + \gamma^2 R_{t+3} + \ldots + \gamma^{T-1} R_T = \sum_{k=0}^{T} \gamma^k R_{t+k+1} \tag{2.22}$$

The γ parameter is the discount rate, it can vary between 0 and 1 and represents the weight of future rewards compared to the actual reward. If γ approximates to zero, future rewards will be minor compared to the current, to the point that, if $\gamma = 0$, only the present reward is taking into account. Whereas if the discount rate is approximated to 1, then future rewards are equally important than the current, such is the case that, if $\gamma = 1$, (2.22) is equal to (2.21) [41].

Depending on the formulation of the RL problem, the structure of the algorithm can formulate in different ways. If the problem needs to be iteratively formulated, the system works in multiples episodes (state intervals between the initial and terminal state), and in that case, before the beginning of a new episode, both time step and return are restored. Otherwise, the problem is formulated to work continuously and the time of execution is infinite; therefore, the return also approaches to infinite.

2.3.2.3 Value Iteration

The value function $v_\pi(s)$ (or v-value) of a state is the expected return starting from the state s and following a policy π is given by

$$v_\pi(s) \doteq \mathbf{E}_\pi\left[G_t | s_t = s\right] = \mathbf{E}_\pi\left[\sum_{k=0}^{\infty} \gamma^k r_{t+k+1} | s_t = s\right], \forall s \in S, \tag{2.23}$$

where $\mathbf{E}_\pi$ refers to the expected value of a random variable given π and the time step t. The value function is a useful variable for the agent to be able to compare between different policies based only on the expected reward and the state where it is, and thus be able to determinate and choose the one that maximizes long-term reward.

In the same way, there is a function that expresses the expected value for an action a and state s, following a policy π, called state-value function (or q-value), expressed in (2.24).

$$q_\pi(s,a) \doteq \mathbf{E}_\pi\left[G_t |, s_t = s |, a_t = a\right] = \mathbf{E}_\pi\left[\sum_{k=0}^{\infty} \gamma^k r_{t+k+1} |, s_t = s |, a_t = a\right] \tag{2.24}$$

To solve an RL problem, an optimal policy must be found. Supposing that the policy π is optimal, then $\pi' \leq \pi$, where π' is any policy different from π. This means that $v_\pi'(s) \leq v_\pi(s)$; therefore, there is an optimal value function and an optimal state-value function, which are defined as follows:

$$v_*(s) \doteq \max v_\pi(s) \tag{2.25}$$

$$q_*(s) \doteq \max q_\pi(s) \tag{2.26}$$

For each state, the value function and the state-value function are estimated by the experience of the agent, taking into account the times the agent visits that state. For this reason, an algorithm is used to converge to the correct v values, known as value iteration.

Value iteration is based on the Bellman optimality equation that specifies that under an ideal policy, the optimal value of a state is the value of an action with the maximum expected return. This means that the optimal action to execute for a given state is the one with the highest q-value.

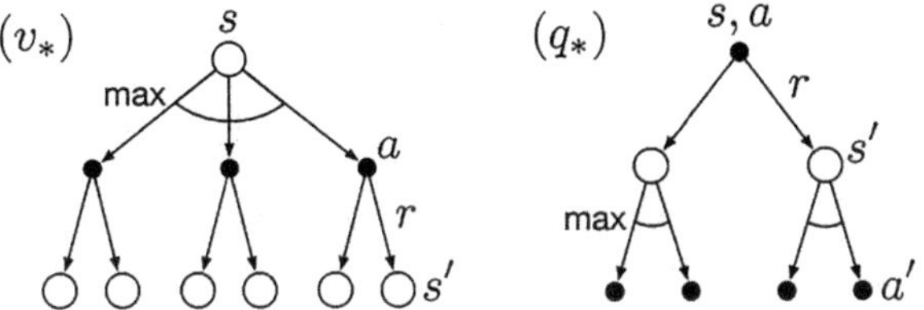

Fig. 2.9 Progressions of agent based on Bellman optimality principle

$$v_*(s) = \max_a \sum_{s',r} p(s', r \mid s, a)\left[r + \gamma v_*(s')\right] \tag{2.27}$$

$$q_*(s) = \sum_{s',r} p(s', r \mid s, a)\left[r + \gamma \max_{a'} q_*(s', a')\right] \tag{2.28}$$

The value iteration algorithm consists of applying (2.27) sweeping the entire state space using a FOR loop. That is repeated for a preset number of iterations or until the variation of the value function is small, as observed in Algorithm 1. The advantage of this one it does not require too many sweeps; hence, it does not need a high computational process to execute it.

Figure 2.9 illustrate progressions of states (white circles) and actions (black circles) considering on Bellman optimality equation for v^* and q^*. The choice of the agent defines the next point from maximum value instead of expected value based on a given policy.

Algorithm 1 – Value iteration
Parameter: $\theta > 0$ determining accuracy of estimation
Initialize $V(s)$ $\forall s \in S$, arbitrarily except that $V(term.) = 0$
Loop for each $s \in S$:
$v \leftarrow V(s)$
$v(s) = \max_a \sum_{s'} p(s' \mid s, a)\left[r + \gamma v(s')\right]$
$\Delta \leftarrow \max(\Delta, \lvert v - V(s)\rvert)$
Until $\Delta < \theta$
Output: deterministic policy $\pi \approx \pi_*$, such that
$\pi(s) = \operatorname{argmax}_a \sum_{s',r} p(s', \lvert, r, \lvert, s, \lvert, a)\left[r + \gamma V(s')\right]$

2.3.2.4 Policy Iteration

As can be seen in (2.27) and (2.28), v^* and q^* depend on values of subsequent states; therefore, in certain occasions, this makes it impossible to obtain values of these functions. In those cases, policy iteration can be used to solve the equations recursively.

The policy iteration is an algorithm that, instead of searching an optimal value function, chooses a policy arbitrarily and manipulates it, updating it iteratively. Considering all states changes and possible actions and selecting in each state the best action according to $q_\pi(s, a)$, for analyzing a new policy π', the following equation is used:

$$\pi'(s) = \underset{a}{\text{argmax}} \sum_{s',r} p(s',|,r,|,s,|,a)\left[r + \gamma v_\pi(s')\right] \tag{2.29}$$

If we suppose that π' is as good as the previous policy, then $v_\pi = v_\pi'$ and (2.29) becomes (2.27), that means that π and π' are optimal policies. The optimal policy is expressed as π^*.

As can be seen in Algorithm 2, the policy iteration is divided into two phases: one that evaluates the policy, by calculating the value function and another that updates said policy to a better one. Through the aforementioned recursion, these stages generate a sequence from the use of the value function to achieve a better policy until reaching a convergence if the number of policies is finite.

Algorithm 2 – Policy iteration
1: Initialize $V(s)$ y π' arbitrarily for all s ϵ S
2: Policy evaluation Loop $\Delta \leftarrow 0$
Loop for each $s \in S$:
$v \leftarrow V(s)$
$V(s) \leftarrow \sum_{s'r} p(s',r \mid s,\pi(s))\left[r + \gamma v(s')\right]$
$\Delta \leftarrow \max(\Delta, \lvert v - V(s) \rvert)$
Until $\Delta < \theta$ (small positive number defining accuracy of estimation)
3: Policy improvement
$policy\_stable \leftarrow true$ For each $s \in S$:
$old\_action \leftarrow \pi(s)$ $\pi(s) \leftarrow \underset{a}{\text{argmax}} \sum_{s',r} p(s',|,r,|,s,|,a)\left[r + \gamma v_\pi(s')\right]$
If $old\_action \neq \pi(s)$, then $policy\_stable \leftarrow false$
If $policy\_stable == true$,
Stop and return $V \approx v_*$ and $\pi \approx \pi_*$; else go to 2

2.3.3 Temporal Difference

Methods based on temporal-difference (TD) learning are part of a larger group of forms of resolution of RL problems, called model-free methods and they are characterized by, as their name indicates, lacking a model that defines the behavior of the agent in its environment [39]. Therefore, it starts from total ignorance of its dynamics, counting only the possible states and actions to be carried out as information.

TD methods have advantages over Monte Carlo and Dynamic Programming methods since they do not require a model of the system under control. For this reason, TD methods are a natural choice for online deployments and in cases where the episode is very slow [42].

The simplest TD method is called TD(0) is known as TD(0) or "one-step TD." There are also two other algorithms based on TD that are very popular: Q-Learning and SARSA. Both have a very similar operating structure, the main difference between them is the use that is given to the learned policy. SARSA is an on-policy method: the agent acts based on the same policy that it is learning, that is, it improves the policy used to make decisions. Instead, Q-Learning is an off-policy method, this means that the evaluated and improved policy is different from the one it uses to choose the actions to be executed. The advantages of one and the other depends on the problem where these methods can be implemented.

2.3.3.1 Q-Learning

Q-Learning (QL) is an off-policy TD-based algorithm that learns from experience generated by direct exploration of the environment. QL policy consists of a tabular form that maps all states with the possible eligible actions by the agent, using respective q-values, see Table 2.1. This table is known as q-table and, usually, is initialized with zeros in each cell and, recursively, the agent updates its q-values.

The agent estimates q-values from the MDP to find an optimal policy, this means that the agent update q-value using the action that has the highest value; however, this action is not always the chosen one in each iteration (exploitation vs. exploration). Using the following equation iteratively, q-value approaches to its ideal value independently of the policy followed:

$$Q(s_t,a_t) \leftarrow Q(s_t,a_t) + \alpha\left[r_{t+1} + \gamma \max_a Q(s_{t+1},a) - Q(s_t,a_t)\right], \tag{2.30}$$

where r_{t+1} is the reward corresponding to the execution of the action a_t and the term $max_a\, Q(s_{t+1}, a)$ represents the ideal q-value for the state s_{t+1}. The α parameter is the learning rate and defines how important is new learning for the update of the new value. If $\alpha = 0$, the system will follow a fixed policy, and new decisions will be discarded, and if $\alpha = 1$, then future decisions will have the greatest possible weight over those already executed. Ideally, the learning rate should be an intermediate value between the limits to ensure the convergence of learning, where the magnitude depends on the problem where the algorithm is implemented.

Algorithm 1 – Episodic Q-Learning
Parameters: step size, $\alpha \in (0,1]$, $\gamma \in [0,1]$, small $\varepsilon > 0$
Initialize $Q(s,a)$, $\forall s \in S\ a \in A(s)$ arbitrarily except that $Q(terminal, \bullet) = 0$ Loop for each episode: Initialize S
Loop for each step of episode:
Choose a from s using policy derived from Q (e.g. ε-greedy)
Take action a, observe r, s'
$Q(s,a) \leftarrow Q(s,a) + \alpha\left[r + \gamma \max_{a} Q(s',a) - Q(s,a)\right]$
$s \leftarrow s'$
until S is terminal

Algorithm 3 is a Q-Learning designed for applications where it is required to work on an episodic basis. It starts from a definition of its parameters and initialization of the q-values, for all states and actions, except the case where S is terminal. The iterative operations are implemented in a double loop, in which the outer loop defines a new episode of the learning task and the inner loop is a step of this episode. In the inner loop, if policy derived is ε-greedy, the agent chooses an action randomly or based on the maximum q-value for the corresponding state. Then, the action is executed and the environment returns a reward signal and the next state, depending on the previous state and action. Finally, the q-value is updated according to (2.30) and the new state becomes the current state, ending the iteration of the inner loop.

For the last state of the inner loop, the number of steps has to be estimated. In the same way, the outer loop must finish when the algorithm achieves a convergent approximation to the optimal policy. If the problem needs to be formulated continuously, then Algorithm N° 3 will be applied only with the inner loop without a specific termination condition.

Table 2.1 Q-table example

		Actions Space A={a1,a2,a3,...am}				
		a1	a2	a3	...	am
States Space S={s1,s2,s3,...,sn}	s1	Q(s1,a1)	Q(s1,a2)	Q(s1,a3)	...	Q(s1,am)
	s2	Q(s2,a1)	Q(s2,a2)	Q(s2,a3)	...	Q(s2,am)
	s3	Q(s3,a1)	Q(s3,a2)	Q(s3,a3)	...	Q(s3,am)
	...	...	...	...	...	...
	sn	Q(sn,a1)	Q(sn,a2)	Q(sn,a3)	...	Q(sn,am)

2.3.3.2 SARSA

Like Q-Learning, the SARSA algorithm is another TD method. Its acronyms come from the words sequence State-Action-Reward-State-Action, due to the sequential procedure to update q-value. Starting from state s, the agent chooses and takes action a, and gets a reward r for such action. Then, it passes to a new state s' and proceeds to execute a new action a'.

The update of the q-value is expressed analytically in (2.31), which recursively leads to convergence of optimal values.

$$Q(s_t,a_t) \leftarrow Q(s_t,a_t) + \alpha\left[r_{t+1} + \gamma \max_a Q(s_{t+1},a_{t+1}) - Q(s_t,a_t)\right] \tag{2.31}$$

Algorithm 4, describes the steps for episodic SARSA

Algorithm 4 – Episodic SARSA

Parameters: step size, $\alpha \in (0, 1]$, $\gamma \in [0, 1]$, small $\varepsilon > 0$

Initialize $Q(s, a)$, $\forall s \in S$ $a \in A(s)$ arbitrarily except that $Q(terminal, \bullet) = 0$

  Loop for each episode:

    Initialize s

    Choose a from s using policy derived from Q (e.g., ε-greedy)

    Loop for each step of episode:

      Take action a, observe r, s'

      Choose a' from s' using policy derived from Q (e.g., ε-greedy)

$$Q(s,a) \leftarrow Q(s,a) + \alpha\left[r + \gamma \max_a Q(s',a') - Q(s,a)\right]$$

      $s \leftarrow s'$, $a \leftarrow a'$

  Until s is terminal

There is an example based on a grid world that emphasizes the main differences between off-policy and on-policy methods, comparing Q-Learning and SARSA. This example is known as *Cliff Walking* and consists of a 4 × 12 position grid, characterized by having a starting cell, a goal, and a cliff that blocks the straight trajectory between these cells, as reflected in Fig. 2.10.

Each action that does not lead to the goal corresponds to a reward equivalent to -1; however, the fall through this abyss generates a very high penalization ($R = -100$). This generates uncertainty between taking the optimal path or the safe path. The optimal path consists of skirting the cliff to reach the objective sooner, at the risk of falling, whereas the safe path is longer avoiding high penalizations for falling off the cliff, performing more actions than the optimal path does.

In Q-Learning (off-policy), the agent assumes that the policy chosen is optimal and will always perform an action based on the maximum q-value, to maximize the long-term reward, except in cases where it decides to randomly explore other possible actions. Hence, it will try to go along the optimal path skirting the cliff and as a consequence of the random exploration, repeatedly fall off the cliff obtaining high

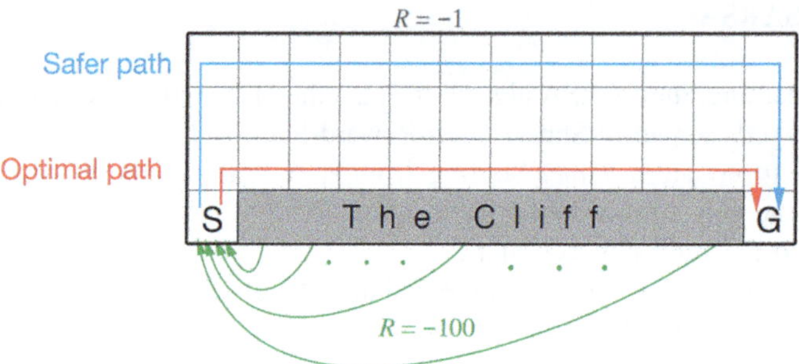

Fig. 2.10 The cliff walking environment

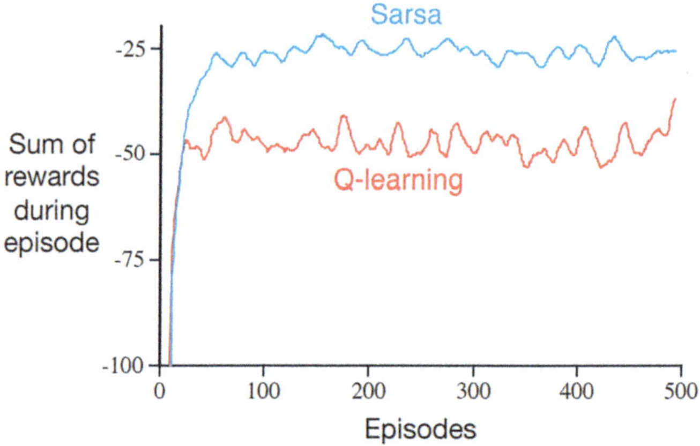

Fig. 2.11 Reward accumulated for episode in function of the number of episodes (R. Sutton, 2018)

penalizations in the accumulated reward. Nevertheless, it does not use information from state-action pairs to update the q-table.

However, SARSA (on-policy) updates the q-table using the state-action function learned from actions chosen by the current policy, meaning the policy that it is learning in that instant. As a result, the agent takes a path further away from the cliff, being more conservative and obtaining more minimum penalizations than Q-Learning, but at the same time, avoiding high penalizations for falling off the cliff.

Figure 2.11 shows reward accumulated during training in function of the number of episodes for $\varepsilon = 0.1$. As appreciated, despite going on the optimal path, the accumulated reward for Q-Learning is less due to the constant falls through the cliff by the ε-greedy policy taken.

From this example, it is demonstrated that depending on the required problem, the implementation of one of the algorithms may be more beneficial than the other.

2.4 Experiments

2.4.1 Problem Formulation

According to the definitions in Sect. 2.3, the interface between agent and environment is given by a simulated model of a single PV panel or module that generates output data from the simulated transformation of received solar energy into electrical energy. At the same time, the controller is responsible for receiving the output data and using it in states and rewards. Furthermore, the controller executes actions that will take the operating point to the MPP, and the maximum power obtained is used by a power converter that adapts the output of the PV module and provides it to a certain load.

The PV panel was modeled using the equations shown in Sect. 2.2, considering variables to the climatic conditions in the time domain. Table 2.2 shows the model parameters, considering a module formed by 36 cells in series.

The model was coded in a function script that requests as inputs: irradiance, module temperature, and voltage, and as outputs: module voltage, current, and power. This function is called during the execution of the learning algorithms.

Firstly, the formulation of the environment for MPPT was made, which works for any RL algorithm that we want to implement.

For state space, it is considered as the states variables: output voltage and power of the panel because these variables are able to define a spatial location on the I-V and P-V curves, and how far the power is from the MPP. As climatic conditions can be variable, curves produced are dynamic in time and as a result, the MPP also is altered under the same domain (showed in Figs. 2.6 and 2.7).

To avoid that the produced state space is infinite, its states are formed solely by combinations of possible voltage and power given by the PV model. For the same purpose, a discretization of the state variables was performed.

Action space is formed by a discrete set of voltage values. According to the equation of electric power, a variation in voltage also creates a variation in power, generating an approach or distancing of the output power of the module in relation to the MPP. For example, under fixed environmental conditions, the actual state is given by voltage V_1 and power $P = P(V_1, I_1)$, where I_1 is the panel current, and the action ΔV_1 is executed, so the next state will be determined by voltage $V_2 = V_1 + \Delta V_1$ and power $P = P(V_2, I_2)$.

The reward for each action executed by the agent depends on the power magnitude of the current step and the variation in power that exists between the current step and the previous one. While current power is positive, meaning, between short-circuit voltage and open-circuit voltage, the agent is rewarded with every approach to the MPP and penalized with every distancing. In the case that the current power is negative, we define a high penalization for distancing and a low penalization for closeness to the MPP.

Table 2.2 Variables of the PV model

Denomination	Magnitude	Unit
k	$1.38 * 10^{-23}$	J/K
q	$1.6 * 10^{-19}$	C
η	1.2	–
V_G	1.12	eV
T_1	298	K
$V_{OC(T1)}$	21.06	V
$I_{SC(T1)}$	3.8	A
T_2	348	K
$I_{SC(T2)}$	3.92	A

2.4.2 Problem Simulation

To solve the formulated problem, Q-Learning and SARSA algorithms were designed, in order to be able to compare and determine which is the most suitable to be implemented through an analysis of their training and validation. Both fulfill the purpose of effect new learning starting from total ignorance of the environment by the agent to finally obtain a q-table after execution of a certain number of episodes.

2.4.2.1 Q-Learning and SARSA

The RL algorithms programmed, according to Algorithm 5 and Algorithm 6, firstly initialize as variables: ranges and resolutions of irradiation, temperature, and current necessary to the generation and discretization of state space and action space. The action space is formed by six different voltage values. In addition, they are initialized initial parameters for learning: discount rate γ, ε-greedy factor, epsilon decay factor, and the total number of episodes and steps.

The epsilon decay factor generates a decrease of the ε-greedy throughout the learning, because its value is less than 1 and multiplies the ε-greedy at each step. The objective is to cause progressively a tendency in the ε-greedy policy to exploit previous experiences over random exploration.

For each episode start, there are defined climatic conditions and state variables, which characterize that start. Climatic conditions remain constant throughout each episode but may or may not be variable during episode transitions, depending on the test performed. Magnitudes are randomly selected from a set predefined by the range and resolutions previously mentioned, allowing the agent to start new episodes under different scenarios. The voltage V_o can acquire any value between 0 and open-circuit voltage V_{oc}.

Previous to the start of the first step, the power value P_0 corresponding to voltage V_0 is calculated by PV function call, and the search of the initial state is performed to start recursive processing within the episode. It is from this point that the differences between Q-Learning and SARSA are observed in the algorithms, concerning the order of execution of operations, as shown in Sects. 2.3.3.1 and 2.3.3.2.

The operations of the inner loop include: choose a ΔV action by ε-greedy policy, perform the action to obtain operating voltage and then the output power by calling PV function, evaluate reward and limits, search new state using the voltage and power obtained, update q-value and variables for a new iteration.

According to formulation considered, if the current step power is positive, the reward R is given by

$$R = 2\Delta P = 2\left(P_{\mathrm{F}} - P_{\mathrm{O}}\right), \tag{2.32}$$

where P_{F} is the current step power and P_0 is the previous step power. In addition, R was defined as a constant value of −50 when P_{F} and ΔP are negative, and a constant value of −5 as a low penalty when P_{F} is not positive and ΔP is positive.

Limits were defined to prevent the agent from trying to leave the state space, by voltage range [−5, 25]. In case of an attempt to exit, automatically the algorithm performs a return to the limit that it crossed.

At the end of each episode, the moving average is calculated with a subset of 200 episodes to smooth curves fluctuations due to a large number of episodes. Finally, at the end of the algorithm, other metrics are calculated to determine the performance of the learning and be used in corresponding figures.

Algorithm 2 - MPPT Q-Learning
Define limits to generation of state space V_{UL}, V_{LL}
Define climatic variables
Discretize state variables
Generate state space $\mathbb{S}$
Define discrete action space $\mathbb{A}$
Define learning parameters: $\alpha, \gamma, \varepsilon, \theta, n^{\circ} episodes, n^{\circ}\ steps$
Initialize $Q(s,a)$, $\forall s \in \mathbb{S}$ $a \in A(s)$ arbitrarily except that $Q(terminal, \bullet) = 0$
For episode $i = 1$ to $n^{\circ} episodes$, do
Define initial conditions of the episode i
Determinate initial state $s(V_O, P_O)$
For step $j = 1$ to $n^{\circ} steps$
Choose action $A = \Delta V$ from space $\mathbb{A}(s)$ using ε-greedy
$V_F \leftarrow V_O + \Delta V$
Call PV model function to obtain P_F and I_F
$\Delta P \leftarrow P_F + P_O$
If $P_F \geq 0$, then $R \leftarrow 2 * \Delta P$
Else if $P_F \leq 0$ and $\Delta P > 0$, then $R \leftarrow -5$

Algorithm 2 - MPPT Q-Learning

Else $R \leftarrow -50$

End If

$R_{accumulated} \leftarrow R_{accumulated} + R * \gamma^j$

If $V_F > V_{UL}$, then $V_F \leftarrow V_{UL}$; End If

If $V_F < V_{LL}$ then $V_F \leftarrow V_{LL}$; End If

Evaluate and search new state $s'(V_F, P_F)$

$$Q(S,A) \leftarrow Q(S,A) + \alpha\left[R + \gamma \max_a Q(S',A) - Q(S,A)\right]$$

$s \leftarrow s'$, $V_O \leftarrow V_F$, $P_O \leftarrow P_F$

$\varepsilon \leftarrow \varepsilon * \theta$

End For

Obtain metrics for each episode

Average rewards

End For

Graphic learning curves

Algorithm 3 - MPPT SARSA

Define limits to generation of state space V_{UL}, V_{LL}

Define climatic variables

Discretize state variables

Generate state space $\mathbb{S}$

Define discrete action space $\mathbb{A}$

Define learning parameters: α, γ, ε, θ, n° *episodes*, n° *steps*

Initialize $Q(s,a)$, $\forall S \in \mathbb{S}$ $A \in \mathbb{A}(S)$ arbitrarily except that $Q(terminal, \bullet) = 0$

For episode $i = 1$ to n° *episodes*, do

Define initial conditions of the episode i

Determinate initial state $S(V_O, P_O)$

Choose action ΔV from space $\mathbb{A}(s)$ using ε-greedy

For step $j = 1$ to n° *steps*

$V_F \leftarrow V_O + \Delta V$

Call PV model function to obtain P_F and I_F

$\Delta P \leftarrow P_F - P_O$

If $P_F \geq 0$, then $R \leftarrow 2 * \Delta P$

Else if $P_F \leq 0$ and $\Delta P > 0$, then $R \leftarrow -5$

Else $R \leftarrow -50$

End If

$R_{accumulated} \leftarrow R_{accumulated} + R * \gamma^j$

If $V_F > V_{UL}$, then $V_F \leftarrow V_{UL}$; End If

If $V_F < V_{LL}$ then $V_F \leftarrow V_{LL}$; End If

Algorithm 3 - MPPT SARSA
Evaluate new state $S'(V_F, P_F)$
Choose action $A' = \Delta V'$ from space $\mathbb{A}(s)$ using ε-greedy
$Q(S,A) \leftarrow Q(S,A) + \alpha\left[R + \gamma \max_a Q(S',A') - Q(S,A)\right]$
$S \leftarrow S'$, $\Delta V \leftarrow \Delta V'$, $V_O \leftarrow V_F$, $P_O \leftarrow P_F$
$\varepsilon \leftarrow \varepsilon * \theta$
End For
Obtain metrics for each episode
Average rewards
End For
Graphic learning curves

2.4.2.2 P&O and IncCond

In addition to Q-Learning and SARSA and to compare the performance of these methods with others commonly used as MPPT, P&O and IncCond were incorporated to the validation of the tests, with a fixed action of 0.25 V. Figures 2.12 and 2.13 show the block diagrams corresponding to the algorithms P&O and IncCond implemented.

2.4.3 Results Analysis

To evaluate the methods developed, we evaluate the results obtained from different scenarios with different conditions of temperature and irradiance.

- Test 1: Constant temperature of 30 °C and variable irradiance for each episode, acquiring any value between 0.1 and 1 KW/m^2 with a resolution of 0.01 KW/m^2. Using these parameters, the family of P-V curves is shown in Fig. 2.14 were calculated by the PV model, where the blue circles correspond to the MPP of such curves.
- Test 2: Variations of temperature between 10 and 45 °C in each episode with a resolution of 1 °C and constant irradiance of 1 KW/m^2. The family of curves is represented according to Fig. 2.15.
- Test 3: Variations of temperature and irradiance under the same limits and resolutions for the previous tests. In Fig. 2.16, the family of P-V curves is observed for the corresponding values.

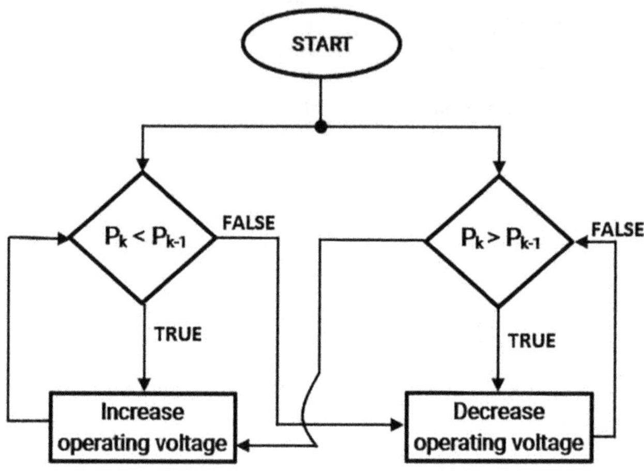

Fig. 2.12 Block diagram for Perturb & Observe algorithm

2.4.3.1 Test 1

The configuration of the environment for this test is observed in Table 2.3, while the agent configuration is shown in Table 2.4. Considering the range and resolution mentioned above, the irradiance was discretized in 91 levels.

Regarding the agent configuration, it was established for training: 25000 episodes formed by 35 steps each, [−5, 25] for voltage range as limits for generation of the state space. The state space is given by a combination of voltage values and resultant power according to the PV model varying all possible voltage and irradiance values within the working range. Taking into account the number of states and the number of actions, a q-table was built with a size of 22,571 rows and 6 columns.

Furthermore, it was defined that a discount rate γ is equal to 0.5, a learning rate equivalent to 0.5, an ε-greedy equal to 1, and an epsilon decay factor θ fixed on 0.999999.

After the training, the corresponding metrics and graphics were obtained. Figure 2.17 corresponds to the comparative chart of moving average of reward accumulated for step with a window size of 200 values in function of the episodes of both learning methods. Approximately, near 10,000 episodes, the rewards are stabilized at a high value per episode, which indicates that the agent has learned to maximize its accumulated reward by acting optimally in each state.

In the case of Fig. 2.18, the moving average (opaque curves) is observed along with the accumulated rewards throughout each iteration per episode (curves with transparency). The randomness of the environmental conditions and the state variables at the beginning of each episode generated the appearance of unknown situations by the agent throughout the training. In consequence, high fluctuations are observed in the curves of the cumulative reward of both methods. Notice that, at the beginning of the episode, the number of steps to reach the MPP is also variable, generating a variation of the value of the cumulative reward. However, as learning

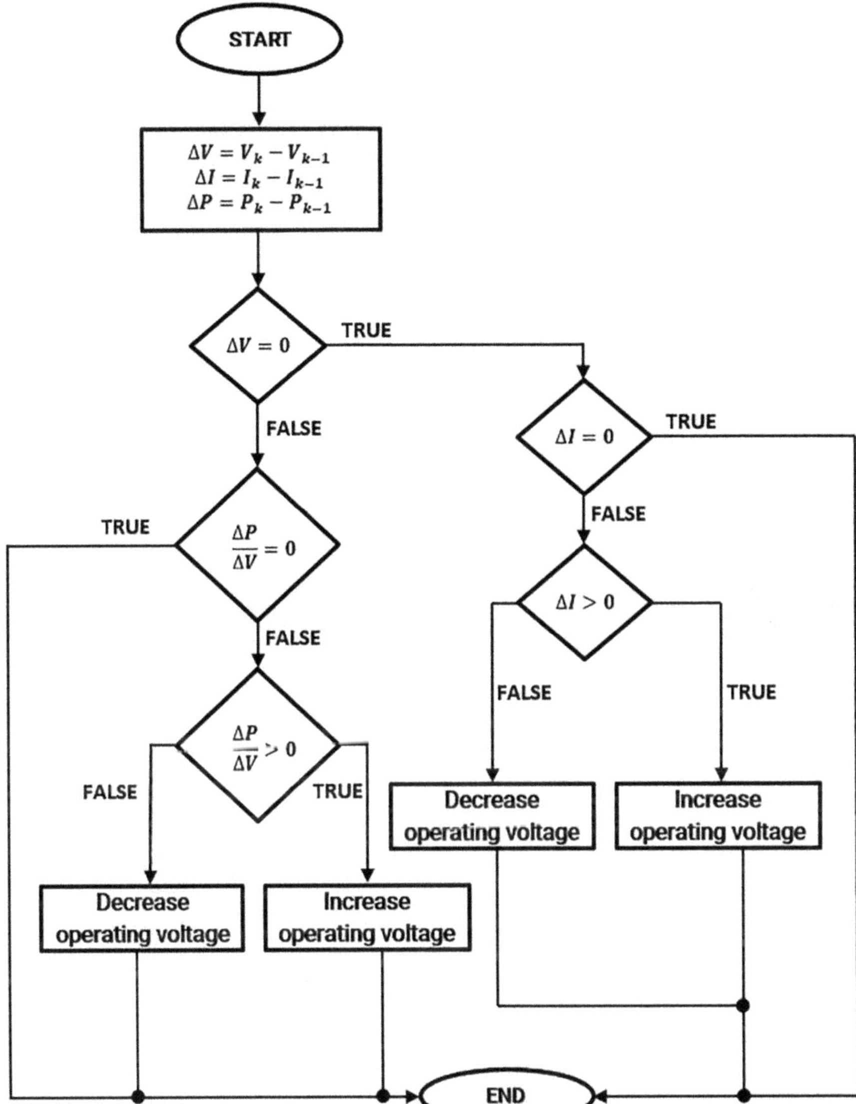

Fig. 2.13 Block diagram for Incremental Conductance algorithm

progresses, it can be seen that the penalizations or negative rewards are decreasing, which means that the agents of both methods learned not to make mistakes.

Table 2.5 shows the mean square error (MSE) and the absolute mean error (MAE) calculated for both techniques taking into account the sum of errors (the difference between real MPP and that obtained by the corresponding method)

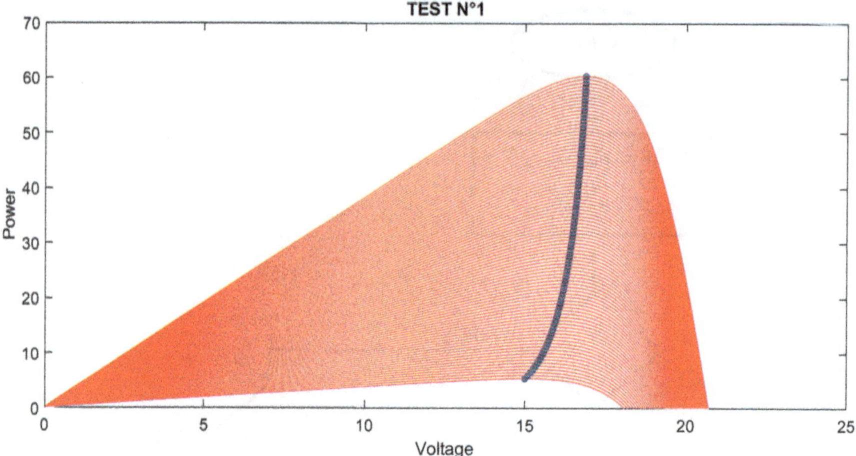

Fig. 2.14 Family of P-V curves corresponding to Test 1

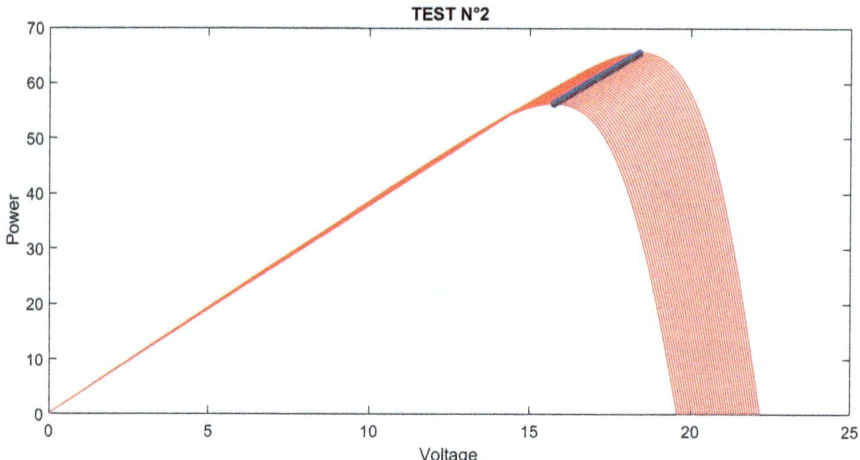

Fig. 2.15 Family of P-V curves corresponding to Test 2

calculated at the end of each episode. SARSA algorithm threw a lower error than Q-Learning during training.

The validation of the trained methods was codified comparing with IncCond and P&O, under the same conditions as those used in training. It consisted of a total of 20,000 episodes formed by 35 steps and in each new episode, initial irradiance and voltage are variable.

Once the validation is finished, the MSE and MAE of each method were calculated by considering the error in each step (second and fourth column) and the error at the end of each episode (third and fifth column), expressed in Table 2.6. Also, curves of power in function to the number of steps were extracted for analysis of the behavior of the most frequent particular cases.

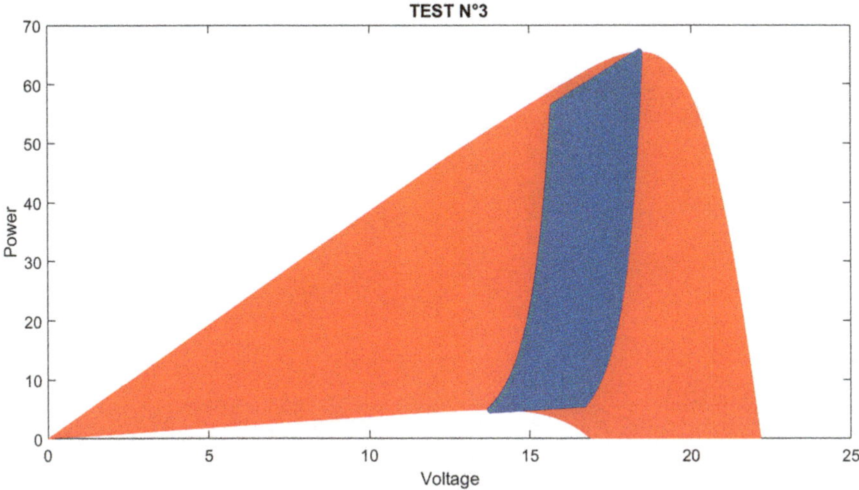

Fig. 2.16 Family of P-V curves corresponding to Test 3

Table 2.3 Variables for configuration of the environment of the Test 1

Variable	Values
Irradiance range	0.1 ~ 1
Irradiance resolution	0.01
Temperature	30

Table 2.4 Variables for configuration of the agent of the Test 1

Variable	Values
State Space	(Voltage, power)
Voltage range	−5 ~ 25
Voltage resolution	0.1 V
Action space	{±3V, ±0.5V, ±0.1V}
Q-table size	27,301 × 6
γ	0.5
α	0.5
ε	1
θ	0.999999
Number of episodes	250,000
Number of steps	35

SARSA presented better metrics throughout the validation process, meaning faster transition speed to approach MPP, as opposed to Q-Learning that demonstrated better convergence to MPP due to better MSE and MAE at the end of each episode.

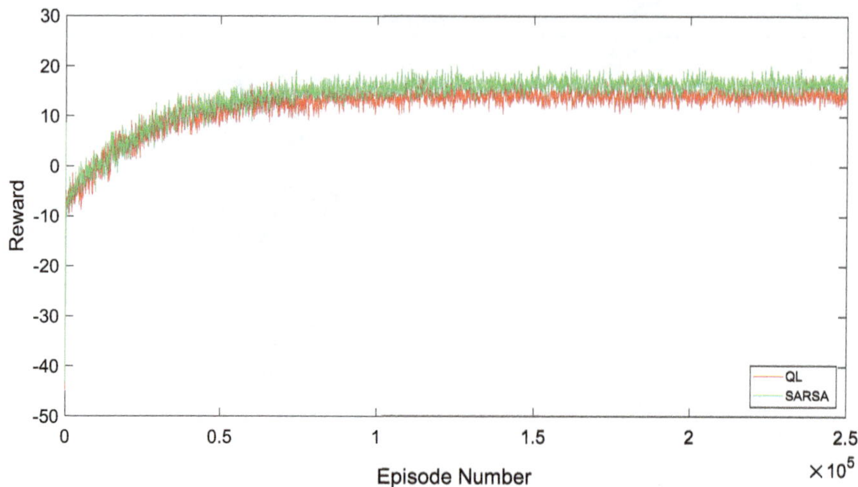

Fig. 2.17 Comparative graph of the moving average of the reward accumulated during the training of Test N ° 1

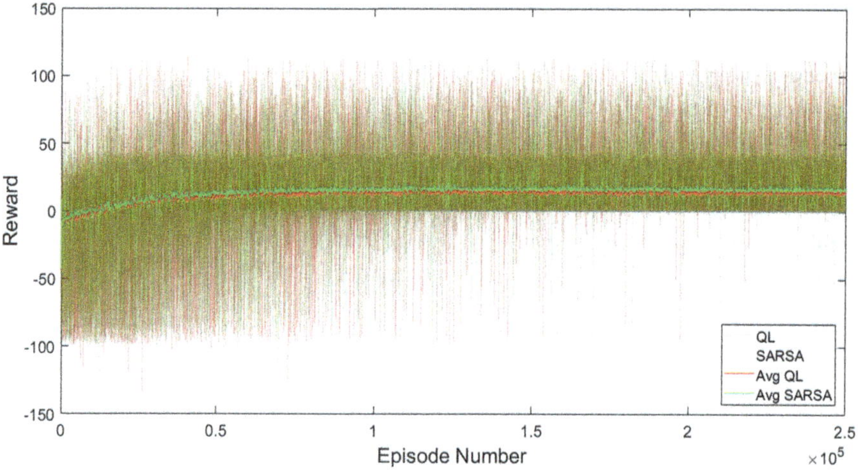

Fig. 2.18 Comparative graph of the reward accumulated per episode during the training of Test N ° 1

Table 2.7 shows episodes randomly obtained from the validation with their results for each algorithm, where *Real MPP* is maximum power points calculated by PV model and *MPP** are those obtained by the models used. Also, *%Error* is the absolute percentage error taking into account only the last step of the corresponding episode. As can be seen, the learning algorithms have better performance than the other algorithms, especially Q-Learning.

Figure 2.19 shows the proximately of the MPPT to the maximum power in function of the number of steps executed with an irradiance of 0.66 KW/m^2. RL methods show a higher precision and stability when compared to traditional methods.

Table 2.5 Variables for configuration of the agent of the Test 1

Method	MSE	MAE
Q-Learning	278.9	3.72
SARSA	280.95	4.06

Table 2.6 Metrics corresponding to validation of the Test 1

Method	MSE (/step)	MSE (/end of ep.)	MAE(/step)	MAE (/end of ep.)
Q-Learning	10.62	0.00003	0.9	0.0033
SARSA	8.77	0.37	0.83	0.36
P&O	146.2	29.07	6.99	2.47
IncCond	146.2	29.07	6.99	2.47

Table 2.7 Results of some episodes of validation of Test 1

N° of	Irradiance.	Real MPP	MPP* (W)			%Error (At end of each ep.)		
Epis.	(KW/m²)	(W)	QL	SARSA	P&O/IC	QL	SARSA	P&O/IC
11	0.66	39.09	39.09	39.09	39.08	0.001%	0.001%	0.003%
8	0.73	43.41	43.41	43.41	39.84	0.021%	0.021%	8.229%
5539	0.51	29.85	29.85	29.47	23.2	0.003%	1.297%	22.28%
924	0.91	54.59	54.56	52.42	52.29	0.042%	3.966%	4.2%
1943	0.54	31.69	31.69	30.17	31.67	0.017%	4.825%	0.093%
16,470	0.54	31.69	31.69	31.51	31.67	0.017%	0.597%	0.093%
13,897	0.25	14.09	14.09	14.09	14.06	0.009%	0.009%	0.265%
6347	0.44	25.57	25.57	25.56	25.53	0.001%	0.043%	0.165%
19,005	0.84	50.24	50.23	49.94	36.32	0.016%	0.578%	27.69%
689	0.45	26.18	26.18	25.77	26.12	0.011%	1.584%	0.23%
8775	0.95	57.08	57.07	56.86	57.06	0.007%	0.381%	0.029%

Additionally, P&O and IncCond have larger fluctuations due to the constant action, which defines their algorithm.

For an irradiance equivalent to 0.92 KW/m^2, as shown in Fig. 2.20, there is a fast approach of the SARSA toward the MPP and to a lesser extent from Q-Learning. Once again, there is a slow response of traditional methods and, therefore, difficulty in stabilizing before ending the episode, generating significant deviation with respect to maximum power at the end of the episode. Consequently, MSE and MAE values or these algorithms are greater than those that correspond to the RL algorithms.

Throughout the validation, P&O and IncCond rarely reached the MPP exactly. These effects are a consequence of the fixed action that is normally defined in these algorithms. If a higher voltage action were defined, it would generate a faster response speed, but the steady-state oscillation would be larger.

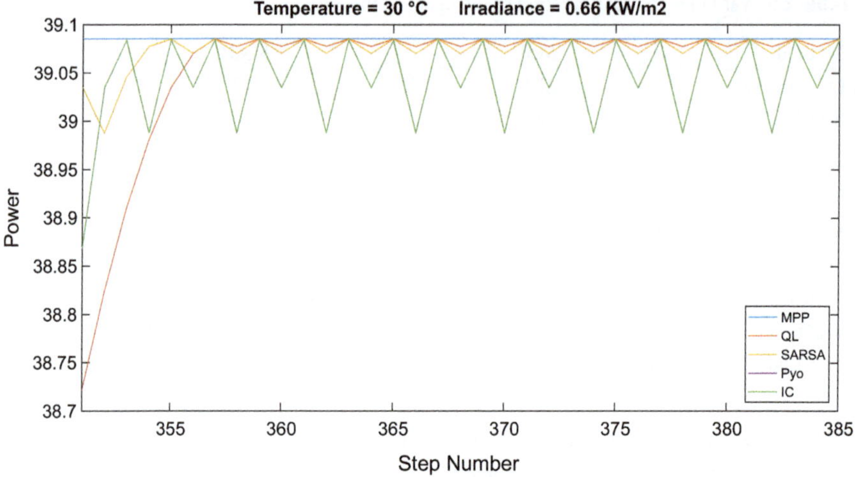

Fig. 2.19 Behavior of MPPT methods in episode 11 of validation of the Test 1

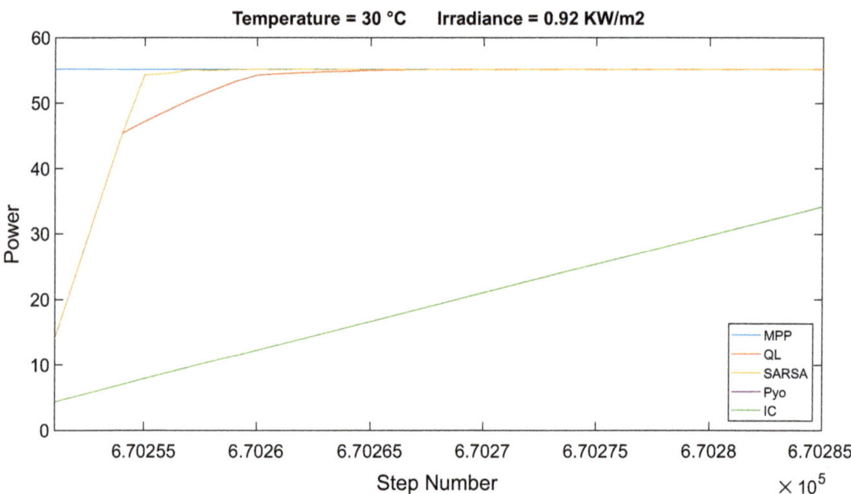

Fig. 2.20 Behavior of MPPT methods in episode 19,151 of validation of the Test 1

2.4.3.2 Test 2

For this test, the environment is configured given the variables observed in Table 2.8. The irradiance was discretized in 36 levels considering the defined range and resolution. The same agent configuration was used, as in Test 1, with the difference of a smaller size of q-table due to conditions used shown in Table 2.9.

Table 2.8 Variables for configuration of the environment of the Test 2

Variable	Values
Temperature range	10 ~ 45
Temperature resolution	1
Irradiance	1

Table 2.9 Variables for configuration of the agent of the Test 2

Variable	Values
State Space	(Voltage, power)
Voltage range	−5 ~ 25
Voltage resolution	0.1 V
Action space	{±3V, ±0.5V, ±0.1V}
Q-table size	10,801 × 6
γ	0.5
α	0.5
ε	1
θ	0.999999
Number of episodes	250,000
Number of steps	35

Once finished the training, the results of the metrics and the corresponding graphs were obtained. According to Table 2.10, training of the Test 2 is quite similar performance to Test 1.

Figure 2.21 shows the balance of cumulative rewards per episode above zero from 150,000 episodes, indicating that they have learned to maximize their reward. The perturbations of the signals are due to the randomness of the temperature and voltage variables that define the beginning of each episode.

As can be seen in Fig. 2.22, the agent makes fewer mistakes as learning progresses; hence, negative rewards disappear in advance phases of learning and positive rewards remain constant.

The validation of this test was performed under the same conditions as in the training, using IncCond and P&O as MPPT methods for comparison. It was comprised of 20 k episodes, each one formed of 35 steps, where the temperature of each episode is a random value within the range expressed in Table 2.8.

Due to that, the short-circuit current keeps constant when only the temperature varies and the initial voltage is zero: the IncCond algorithm, because of its operating methodology, has the disadvantage of keeps static at the beginning of the validation. Consequently, it requires a manual start initially defining a ΔI value of Fig. 2.13 other than zero.

Once the validation was completed, corresponding values of MSE and MAE for each method were calculated, expressed in Table 2.11. Q-Learning has less error than the other methods, as a result of better precision in reaching the MPP. Also, some power curves depending on the number of total steps were obtained.

Table 2.10 Corresponding metrics to the training of Test 2

Method	MSE	MAE
Q-Learning	489.67	5.65
SARSA	492.66	5.91

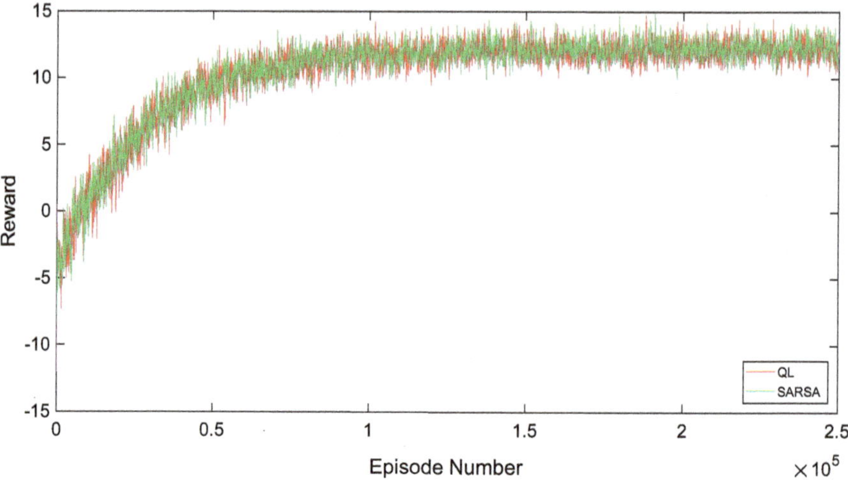

Fig. 2.21 Comparative graph of the moving average of the reward accumulated during the training of Test 2

As observed in Fig. 2.23, when the difference between maximum power and initial power is very large, the number of steps that forms each episode is not enough for the P&O and IncCond methods to converge toward the MPP. Instead, RL methods need few steps to converge to maximum power.

Finally, there are cases as shown in Fig. 2.24, where SARSA response has greater oscillation and error than those of the other systems.

Table 2.12 shows episodes extracted randomly from the validation of this test. Q-Learning demonstrates better performance in most cases.

Through the graphs and metrics shown above, it was concluded that, for these conditions, in general, both RL methods have higher response speed and precision to achieve the MPP, principally Q-Learning for its stability and minimal variations in the order of hundredths.

2.4.3.3 Test 3

Due to variations of temperature and irradiance, and according to the Markov property (expressed in Sect. 2.3.2), for this test, it is not possible to work only with the output voltage and power of the PV panel as input variables. Consequently, for this case, it was defined to add irradiance as a state variable together with the previous

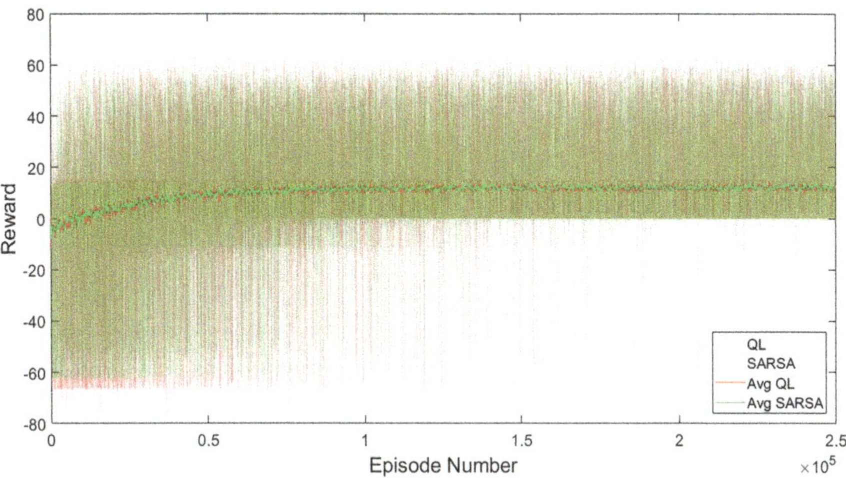

Fig. 2.22 Comparative graph of the reward accumulated per episode during the training of Test 2

Table 2.11 Corresponding metrics to the validation of Test 2

Method	MSE (/step)	MSE (/end of ep.)	MAE(/step)	MAE (/end of ep.)
Q-Learning	24.93	0.00006	1.1	0.006
SARSA	21.21	0.14	1.14	0.23
P&O	416.01	90.29	13.18	4.93
IncCond	416.01	90.29	13.18	4.93

ones mentioned. Below, in Tables 2.13 and 2.14, the configuration that defines environment and agent is defined, respectively.

Table 2.15 shows the MSE and MAE corresponding to the training of this test. According to values obtained in both algorithms, the behavior of the methods during learning was similar.

The comparison of the learning evolution of the RL methods (Fig. 2.25) shows approximately from episode 10,000, a stabilization of cumulative reward per episode above zero, indicating both systems learned to maximize their rewards. Also, in Fig. 2.26, it is observed that the penalties obtained in Q-Learning disappear approximately in the middle of the training, hence, the agent learns not to make mistakes that lead to such negative rewards. In SARSA, the same happens for high penalties, but it is observed that it continues to receive low penalties for moving away from the MPP.

The validation consisted of 50,000 episodes, each formed by 35 steps, under the same conditions as during training. According to the MSE and MAE obtained and expressed in Table 2.16, RL methods show better accuracy in the approach to MPP. In the case of SARSA, it shows better MSE and MAE if the error in each step is taken into account, indicating a faster convergence in comparison to the other

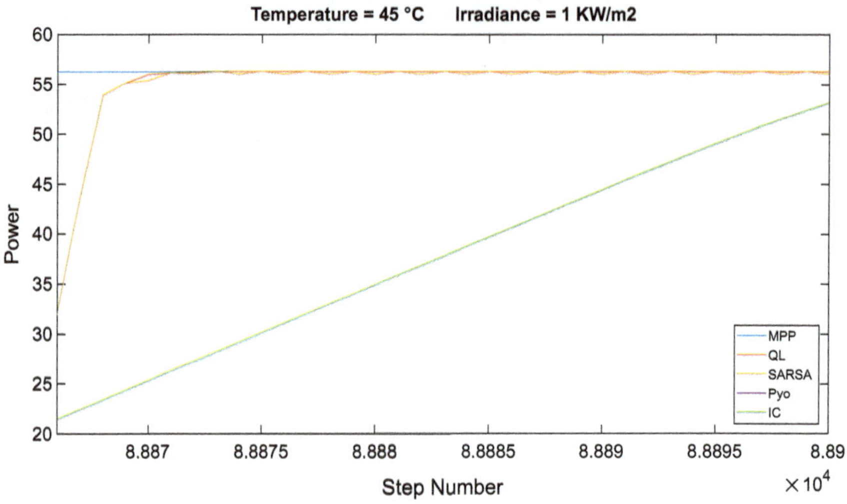

Fig. 2.23 Behavior of MPPT methods in episode 2540 of validation of the Test 2

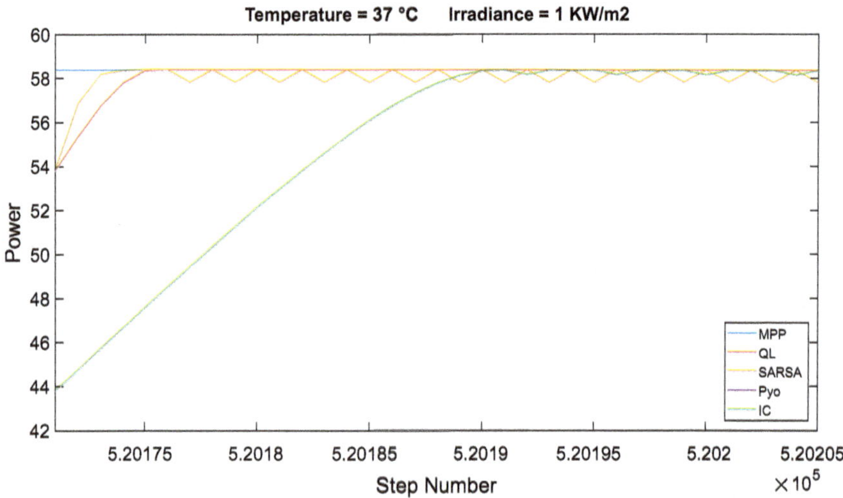

Fig. 2.24 Behavior of MPPT methods in episode 14,863 of validation of the Test 2

systems. Instead, Q-Learning presents a better approximation to the maximum power, since it has a lower MSE and MAE taking into account only final steps.

According to Table 2.17, which shows some episodes randomly extracted from validation with their respective data, generally in cases where the P&O and IC algorithms have enough time to reach the maximum power, the percentage error of these two methods taken at the end of each episode is at a similar level than that corresponding to Q-Learning and above SARSA. However, in cases where they do not reach that level, the percentage error is very high, in some cases over 40%.

Table 2.12 Results of some episodes of validation of Test 2

N° of	Temp.	Real MPP	MPP* (W)			%Error (At the end of each episode)		
Episode	(°C)	(W)	QL	SARSA	P&O/IC	QL	SARSA	P&O/IC
1	39	57.81	57.79	57.79	57.71	0.027%	0.027%	0.166%
2	20	62.81	62.81	62.69	48.26	0.01%	0.201%	23.16%
8	42	57.01	57.01	56.81	56.99	0.002%	0.363%	0.042%
15,904	39	57.81	57.81	57.81	57.81	0%	0%	0.007%
3738	29	60.45	60.45	60.41	59.57	0.009%	0.065%	1.468%
9796	19	63.08	63.06	63.01	63.07	0.02%	0.098%	0.012%
8912	33	59.40	59.39	59.36	52.94	0.008%	0.063%	10.87%
12,927	10	65.42	65.40	65.25	65.38	0.029%	0.262%	0.062%
14,188	37	58.34	58.34	57.76	58.28	0.007%	1.001%	0.108%
15,094	13	64.64	64.63	64.42	33.37	0.015%	0.344%	48.36%
5521	38	58.07	58.07	57.86	58	0.016%	0.363%	0.136%

Table 2.13 Variables for configuration of the environment of the Test 3

VARIABLE	VALUE
Temperature range	10 ~ 45
Temperature resolution	1
Irradiance range	0.1 ~ 1
Irradiance resolution	0.01

Table 2.14 Variables for configuration of the agent of the Test 3

VARIABLE	VALOR
State space	(Irradiance, voltage, power)
Voltage range	−5 ~ 25
Voltage resolution	0.1 V
Action space	{±3V, ±0.5V, ±0.1V}
Q-table size	982,891 × 6
γ	0.5
α	0.5
ε	1
θ	0.999999
Number of episodes	500,000
Number of steps	35

In addition, in the curves of power in function of the number of total steps corresponding to validation, the slow convergence speed of the traditional MPPT methods is reflected before an action of 0.25 V. SARSA algorithm is faster to converge to maximum power, but, in many cases, there is a significant steady-state error, unlike Q-Learning, which is slower but has better accuracy, as shown, for example, in Fig. 2.27.

Table 2.15 Corresponding metrics to the training of Test 3

Method	MSE	MAE
Q-Learning	139.09	2.01
SARSA	139.48	2.55

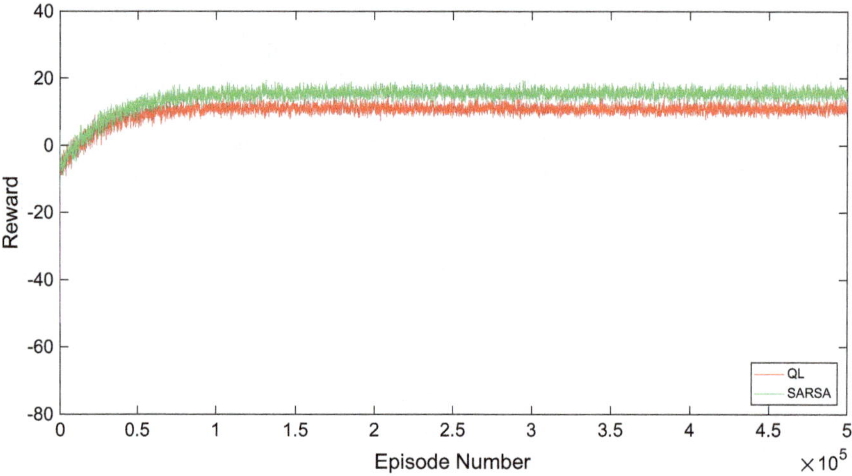

Fig. 2.25 Comparative graph of the moving average of the reward accumulated during the training of Test 3

Table 2.17 Results of some episodes of validation of Test 3

N°. of	Temp.	Irrad.	Real MPP	MPP* (W)			%Error (At the end of each episode)		
Epis.	(°C)	(KW/m^2)	(W)	QL	S	PO/IC	QL	S	PO/IC
8131	42	0.56	31.10	31.09	30.64	31.02	0.048	1.479	0.243
5950	14	0.39	24.23	24.2	24.23	24.23	0.140	0.022	0.004
24,919	24	0.18	10.25	9.85	10.23	10.25	3.927	0.221	0
47,988	23	0.95	58.83	58.61	57.82	58.80	0.366	1.706	0.047
29,264	27	0.95	57.83	57.62	55.14	33.43	0.361	4.643	42.185
11,191	22	0.11	6.14	5.45	6.14	4.56	11.234	0.009	25.676
37,564	43	0.58	32.11	32.11	32.02	32	0.017	0.293	0.346
12,755	24	0.99	61.13	60.96	59.52	35.90	0.276	2.626	41.268
25,298	11	0.81	52.43	52.37	52.11	52.26	0.115	0.621	0.323
34,954	11	0.37	23.23	23.10	22.90	18.59	0.569	1.423	20.008
44,546	41	0.48	26.58	26.55	25.56	26.57	0.087	3.836	0.001

For irradiance values less than 0.2 KW/m^2, Q-Learning shows significant instability (as, e.g., in Fig. 2.28.). For irradiance values greater than 0.8 KW/m^2, the convergence of the system raises significantly as reflected in Fig. 2.29. The variety of actions that the agent can choose and the experience acquired during training

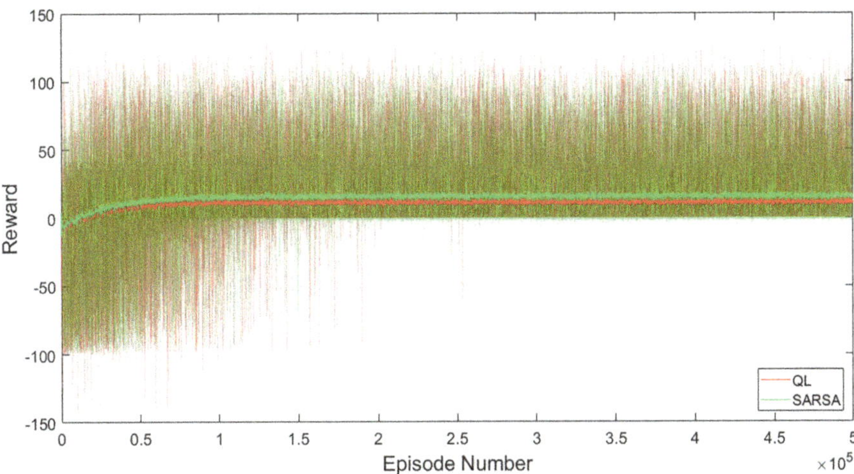

Fig. 2.26 Comparative graph of the reward accumulated per episode during the training of Test 3

Table 2.16 Corresponding metrics to the validation of Test 3

Method	MSE (/step)	MSE (/end of ep.)	MAE(/step)	MAE (/end of ep.)
Q-Learning	23.21	0.077	1.97	0.15
SARSA	11.17	2.09	1.29	0.77
P&O	147.67	30.74	6.99	2.53
IC	147.67	30.74	6.99	2.53

allow RL methods to have better precision and speed of convergence as MPPT is faced with alterations in climatic variables on the PV module.

2.5 Conclusions and Final Remarks

This work presents the development of reinforcement learning control algorithms Q-Learning and SARSA for MPPT control in PV panels under climatic variability conditions. The implemented algorithms were tested in a simulated environment under different operating conditions to evaluate their performance. The results of the tests show an acceptable performance of the algorithms based on RL to achieve efficient MPPT control. In general, they show better performance than traditional and specific techniques such as P&O and IC.

RL techniques have the ability to learn an efficient policy from scratch only by interacting with the controlled PV system. Also, it does not require any a priori knowledge of the PV panel model, its parameters, or additional configurations. Once the optimal policy has been found, it does not require additional computation

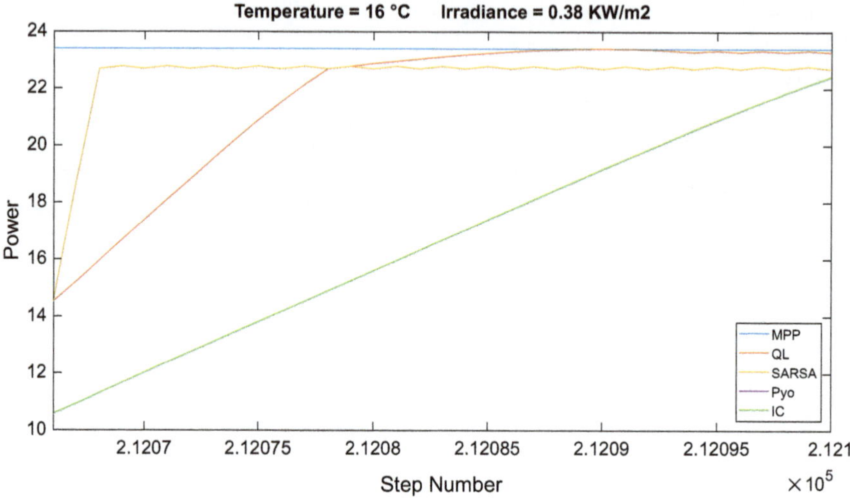

Fig. 2.27 Behavior of MPPT methods in episode 6060 of validation of the Test 3

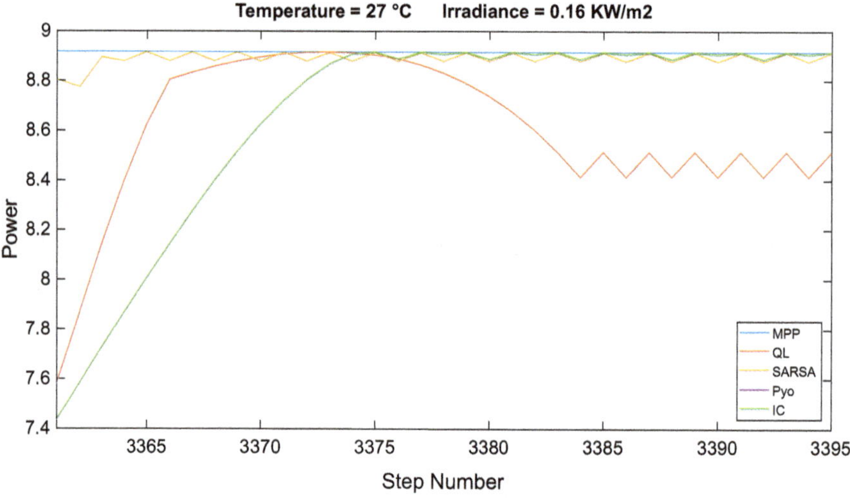

Fig. 2.28 Behavior of MPPT methods in episode 97 of validation of the Test 3

to determine the corresponding output action. Significantly, the optimal policy contemplates variability of the environment and uncertainty on the parameters model.

Nevertheless, accuracy in determining the MPP is achieved at a high computational cost due to the large size of the state space, this limitation results are evident in Test 3 when adding the irradiance value to the input vector state. Deep learning models will be considered aiming to reduce the computational cost of building the state action table. Deep Q-Network, for example, uses neural networks to

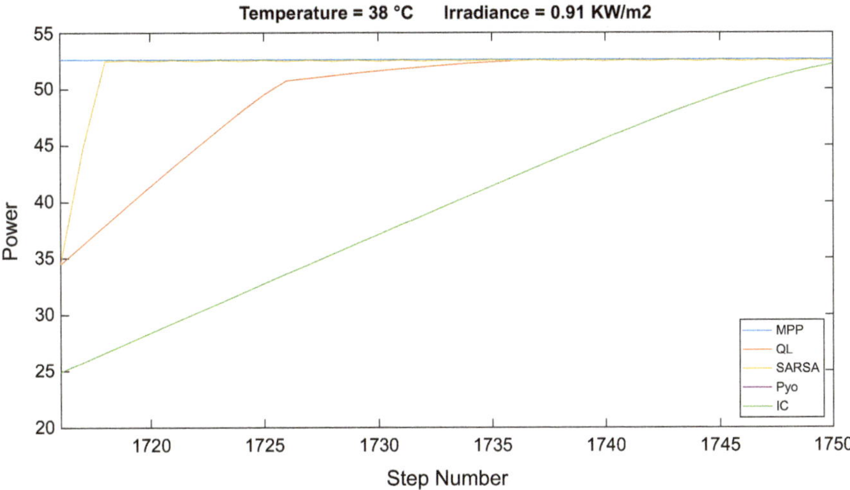

Fig. 2.29 Behavior of MPPT methods in episode 50 of validation of the Test 3

approximate an optimal value function for actions. This has the advantage of no longer requiring to discretize the states and, therefore, the fact of working with continuous spaces makes it possible to maximize the efficiency of the MPPT control.

An oscillation over the MPP is presented in some experiments, since a no operate action was not defined within the space of actions in order to remain on the MPP once it was reached. In order to eliminate the oscillation of control over the MPP, in the short term, zero will be added as an action within its space. Thus, when reaching the maximum power, the algorithms will remain at that value, managing even higher accuracy.

Tests were performed in a simulated environment, the next step involves performing experimental tests in real PV panels. On the other hand, the influence of shading areas caused by clouds, for example, might be incorporated into the optimization problem. On large PV systems, frequently, this situation generates partial shading forming conditions of nonuniform irradiance, which generates an alteration in the forms of the P-V curves shown in the present work that can cause that the traditional MPPT cannot tackle the problem.

References

1. R.C. Hsu, C.T. Liu, W.Y. Chen, H.I. Hsieh, H.L. Wang, A reinforcement learning-based maximum power point tracking method for photovoltaic array. Int. J. Photoenergy **2015** (2015)
2. N.L. Panwar, S.C. Kaushik, S. Kothari, Role of renewable energy sources in environmental protection: A review. Renew. Sust. Energ. Rev. **15**(3), 1513–1524 (2011)
3. H. Islam et al., Performance evaluation of maximum power point tracking approaches and photovoltaic systems. Energies **11**(2), 365 (2018)

4. M. Rahmani-Andebili, Dynamic and adaptive reconfiguration of electrical distribution system including renewables applying stochastic model predictive control. IET Gener. Transm. Distrib. **11**(16), 3912–3921 (2017)
5. M. Rahmani-Andebili, Cooperative distributed energy scheduling in microgrids, in *Power Systems*, no. 9789811070006, (Springer, Singapore, 2018), pp. 235–254
6. M. Rahmani-Andebili, Stochastic, adaptive, and dynamic control of energy storage systems integrated with renewable energy sources for power loss minimization. Renew. Energy **113**, 1462–1471 (2017)
7. D. Baimel, S. Tapuchi, Y. Levron, J. Belikov, Improved fractional open circuit voltage MPPT methods for PV systems. Electron **8**(3), 321 (2019)
8. S. Hadji, J.P. Gaubert, F. Krim, Maximum Power Point Tracking (MPPT) for Photovoltaic systems using open circuit voltage and short circuit current, in 2013 3rd Int. Conf. Syst. Control. ICSC 2013, pp. 87–92, 2013
9. K.M. Tsang, W.L. Chan, Model based rapid maximum power point tracking for photovoltaic systems. Energy Convers. Manag. **70**, 83–89 (2013)
10. P. Bhatnagar, R.K. Nema, Maximum power point tracking control techniques: State-of-the-art in photovoltaic applications. Renew. Sust. Energ. Rev. **23**, 224–241 (2013)
11. D.P. Hohm, M.E. Ropp, Comparative study of maximum power point tracking algorithms. Prog. Photovolt. Res. Appl. **11**(1), 47–62 (2003)
12. P. Jain, S.N. Joshi, N. Gupta, K.G. Sharma, Analysis of MPPT techniques in grid connected PV system, in 3rd Int. Conf. Work. Recent Adv. Innov. Eng. ICRAIE 2018, July 2018
13. R.I. Putri, S. Wibowo, M. Rifa'i, Maximum power point tracking for photovoltaic using incremental conductance method. Energy Procedia **68**, 22–30 (2015)
14. K. Ishaque, Z. Salam, G. Lauss, The performance of perturb and observe and incremental conductance maximum power point tracking method under dynamic weather conditions. Appl. Energy **119**, 228–236 (2014)
15. A.I. Dounis, P. Kofinas, C. Alafodimos, D. Tseles, Adaptive fuzzy gain scheduling PID controller for maximum power point tracking of photovoltaic system. Renew. Energy **60**, 202–214 (2013)
16. K. Loukil, H. Abbes, H. Abid, M. Abid, A. Toumi, Design and implementation of reconfigurable MPPT fuzzy controller for photovoltaic systems. Ain Shams Eng. J. **11**(2), 319–328 (2020)
17. A.I. Dounis, P. Kofinas, G. Papadakis, C. Alafodimos, A direct adaptive neural control for maximum power point tracking of photovoltaic system. Sol. Energy **115**, 145–165 (2015)
18. M. Arjun, J.B. Zubin, Artificial neural network based hybrid MPPT for photovoltaic modules, in 2018 Int. CET Conf. Control. Commun. Comput. IC4 2018, pp. 140–145, Nov. (2018)
19. M.A. Sasi, Fuzzy logic control of MPPT controller for PV systems, Memorial University of Newfoundland, (2017)
20. M. Yaichi, M.K. Fellah, A. Mammeri, A neural network based MPPT technique controller for photovoltaic pumping system. Int. J. Power Electron. Drive Syst. **4**(2), 241–255 (2014)
21. C.C. Aggarwal, *Neural Networks and Deep Learning* (Springer International Publishing, Cham, Switzerland) (2018)
22. P. Kofinas, S. Doltsinis, A.I. Dounis, G.A. Vouros, A reinforcement learning approach for MPPT control method of photovoltaic sources. Renew. Energy **108**, 461–473 (2017)
23. R. S. Sutton and A. G. Barto, Reinforcement Learning: An Introduction Second Edition, in Progress, Second Edi. Cambridge, MA: The MIT Press, (2014)
24. L. De Bernardez, R.H. Buitrago, M. Battioni, M. Cutrera, G. Risso, B. Gottlieb, MEDICIÓN DE LAS CURVAS I-V A OSCURAS DE LAS CELDAS DE UN MODULO FOTOVOLTAICO. Av. en Energías Renov. y Medio Ambient. **8**, 8–12 (2004)
25. M.G. Molina, P.E. Mercado, P.E. Wiernes, Análisis y simulación de algoritmos de control para el seguimiento del punto de máxima potencia de sistemas solares fotovoltaicos conectados a la red eléctrica. Av. en Energías Renov. y Medio Ambient. **11**, 153–160 (2007)
26. T. Ma, H. Yang, L. Lu, Solar photovoltaic system modeling and performance prediction. Renew. Sust. Energ. Rev. **36**. Elsevier Ltd, 304–315 (2014)

27. G. Ciulla, V. Lo Brano, V. Di Dio, G. Cipriani, A comparison of different one-diode models for the representation of I-V characteristic of a PV cell. Renew. Sust. Energ. Rev. **32**, 684–696 (2014)
28. G.R. Walker, Evaluating MPPT converter topologies using a MATLAB PV model, *Australas. Univ. Power Eng. Conf. AUPEC'00*, 2000
29. K.Y. Chou, S.T. Yang, C.S. Yang, Y.P. Chen, Maximum power point tracking of photovoltaic system based on reinforcement learning, in *2019 IEEE Int. Conf. Consum. Electron. - Taiwan, ICCE-TW 2019*, (2019)
30. M. Hlaili, H. Mechergui, Comparison of different MPPT algorithms with a proposed one using a power estimator for grid connected PV systems. Int. J. Photoenergy **2016** (2016)
31. R. Garner, Solar Irradiance, 2015
32. L. Avila, M. De Paula, M. Trimboli, I. Carlucho, Deep reinforcement learning approach for MPPT control of partially shaded PV systems in Smart Grids. Appl. Soft Comput. **97**(xxxx), 106711 (2020)
33. L. Avila, M. De Paula, I. Carlucho, C. Sanchez Reinoso, MPPT for PV systems using deep reinforcement learning algorithms. IEEE Lat. Am. Trans. **17**(12), 2020–2027 (2019)
34. B. Bendib, H. Belmili, F. Krim, A survey of the most used MPPT methods: Conventional and advanced algorithms applied for photovoltaic systems. Renew. Sust. Energ. Rev. **45**. Elsevier Ltd, 637–648 (2015)
35. M.A.G. De Brito, L.P. Sampaio, G. Luigi, G.A.E. Melo, C.A. Canesin, Comparative analysis of MPPT techniques for PV applications, in *3rd International Conference on Clean Electrical Power: Renewable Energy Resources Impact, ICCEP 2011*, (2011), pp. 99–104
36. M.A. Eltawil, Z. Zhao, MPPT techniques for photovoltaic applications. Renew. Sust. Energ. Rev. **25**. Elsevier Ltd, 793–813 (2013)
37. M. Rosu-Hamzescu, S. Oprea, Microchip Technology Inc., Practical guide to implementing solar panel MPPT algorithm, 2013
38. A. Safari, S. Mekhilef, Simulation and hardware implementation of incremental conductance MPPT with direct control method using cuk converter. IEEE Trans. Ind. Electron. **58**(4), 1154–1161 (2011)
39. K.A. Aganah, A.W. Leedy, A constant voltage maximum power point tracking method for solar powered systems, in *Proc. Annu. Southeast. Symp. Syst. Theory*, (March 2011), pp. 125–130
40. L.P. Kaelbling, M.L. Littman, A.W. Moore, Reinforcement learning: A survey. *Journal of artificial intelligence research, 4*, 237–285 (1996)
41. X. Li, Z. Lv, S. Wang, Z. Wei, L. Wu, A reinforcement learning model based on temporal difference algorithm. IEEE Access **7**, 121922–121930 (2019)
42. L. Bom, R. Henken, M. Wiering, Reinforcement learning to train Ms. Pac-Man using higher-order action-relative inputs, in *IEEE Symposium on Adaptive Dynamic Programming and Reinforcement Learning, ADPRL*, (2013), pp. 156–163
43. J. del R. Millán, "Reinforcement learning of goal-directed obstacle-avoiding reaction strategies in an autonomous mobile robot," Robot. Auton. Syst., vol. 15, no. 4, pp. 275–299, Oct. 1995
44. S.Y. Oh, J.H. Lee, D.H. Choi, A new reinforcement learning vehicle control architecture for vision-based road following. IEEE Trans. Veh. Technol. **49**(3), 997–1005 (2000)
45. C. Wei, Z. Zhang, W. Qiao, L. Qu, Reinforcement-learning-based intelligent maximum power point tracking control for wind energy conversion systems. IEEE Trans. Ind. Electron. **62**(10), 6360–6370 (2015)
46. D.P. Bertsekas, J.N. Tsitsiklis, Introduction to probability, Cambridge, MA
47. S.J. Gershman, N.D. Daw, Reinforcement learning and episodic memory in humans and animals: An integrative framework. Annu. Rev. Psychol. **68**(1), 101–128 (2017)

Chapter 3
A Novel Three-Stage Short-Term Photovoltaic Prediction Approach Based on Neighborhood Component Analysis and ANN Optimized with PSO (NCA-PSO-ANN)

Eric Ofori-Ntow Jnr, Yao Yevenyo Ziggah, Mehdi Rahmani-Andebili, Maria Joao Rodrigues, and Susana Relvas

Abstract Parameters of solar radiation are unstable, which affect the prediction accuracy of photovoltaic (PV) power. Many hybrid models have been applied recently to improve the prediction accuracy of the models. In this chapter, a novel three-stage hybrid model of Neighborhood Component Analysis (NCA), and Artificial Neuronal Network (ANN) optimized with Particle Swarm Optimization (PSO) is introduced. First, the NCA technique is applied as a feature extraction technique to determine the relevant features that have substantial influence on photovoltaic power. The study now applies the chosen feature components as inputs into the optimized ANN with PSO to create the PV prediction model. The prediction

E. Ofori-Ntow Jnr (✉)
CEG-IST, Instituto Superior Tecnico, Universidade de Lisboa, Lisbon, Portugal

Faculty of Engineering, University of Mines and Technology, Tarkwa, Ghana
e-mail: eric.jnr@tecnico.ulisboa.pt; eofori@umat.edu.gh

Y. Y. Ziggah
Faculty of Geosciences and Environmental Studies, University of Mines and Technology, Tarkwa, Ghana
e-mail: yyziggah@umat.edu.gh

M. Rahmani-Andebili
Electrical Engineering Department, Montana Technological University, Butte, MT, USA
e-mail: mrahmaniandebili@mtech.edu

M. J. Rodrigues
Lisboa E-Nova and Instituto Superior Tecnico, Universidade de Lisboa, Lisbon, Portugal
e-mail: mariarodrigues@lisboaenova.org

S. Relvas
CEG-IST, Instituto Superior Tecnico, Universidade de Lisboa, Lisbon, Portugal
e-mail: susana.relvas@tecnico.ulisboa.pt

M. Rahmani-Andebili (ed.), *Applications of Artificial Intelligence in Planning and Operation of Smart Grids*, Power Systems,
https://doi.org/10.1007/978-3-030-94522-0_3

strength of the proposed NCA-PSO-ANN model is compared with other variant hybrid models such as Principal Component Analysis (PCA), PSO and ANN (PCA-PSO-ANN), PSO-ANN, and PCA-ANN. The proposed NCA-PSO-ANN model constitutes a very reliable computational tool as it performed better in the selected performance indexes than the compared models.

Keywords Neighborhood Component Analysis · Artificial neural network · Dimensionality reduction · short-term forecasting

3.1 Introduction

Current desire of shedding off fossil fuel powered energy generation to adopt renewable energy power source is getting more popular because of the environmental issues associated with fossil fuel power generation plants. Integrating renewable energy onto grid is in high demand recently, especially in the developing world [1]. Photovoltaic (PV) power, which is one of the potential renewable energy sources, is the most abundantly available energy source, very cheap, and highly accessible [2]. Cost associated with fuel powered energy generation has made many developing countries to now shift their attention toward PV power source at a faster rate [3]. It is established that the peak of PV power coincides with the period when energy demand is also at its peak, making PV power more advantageous. It is again expected that future PV installations will exceed hundreds of Megawatts in terms of power generation capacity [3–5]. A major drawback associated with PV power source is the frequent interruptions in its availability. Addressing this drawback calls for efficient forecasting methods [6]. Many studies have therefore discussed the issue of variability in the PV power and proposed more efficient forecasting methods to be able to harness the full potentials associated with PV power [7]. One of the techniques to counter the problem of variability is to develop a better short-term forecasting method for the purpose of adequate planning and harnessing all the quality power produced from PV source. There is also the need to study the behavior of renewable energy to come out with a well-planned strategy to boost the benefits of integrating it into grid [8]. In view of the foregoing discussion, the most relevant research works on this subject matter are presented.

Mellit et al. [9] used deep learning neural network to forecast short-term PV power using PV data from University of Trieste in Italy. They considered four different time horizons using one-step and multisteps ahead to forecast PV power. Their study produced a very high correlation coefficient, which indicates that there is a significant agreement between the actual and forecasted PV values. Abdel-Basset et al. [10] employed a new deep learning approach called PV-Net to forecast short-term PV power utilizing a real world data sourced from Alice Spring in Australia. The study pointed out some limitations considering the inability of the study to consider some relevant factors that affect PV power production. The authors concluded that when these factors are considered, the results may be affected. Liu et al. [11] used a new hybrid ensemble empirical mode decomposition (EEMD) with extreme learning machine (ELM) optimized by grey wolf optimization (GWO) to forecast the PV power. The model converges at high accuracy and has a good stability but with some limitations. The model did not consider boundary effect, which is important when decomposing the mode of PV. A study done by Hossain and Mahmood [12] proposed a long short-term memory neural network to test single and rolling horizon for forecasting PV power. The result was compared with nonlinear autoregressive exogenous neural network and generalized recurrent neural network. The data used was grouped into different weather seasons (winter, spring, summer, and fall). The used variables included precipitable water, solar irradiance, pressure, wind speed, and relative humility. The authors proved in their work the negative effect of increasing the input features when predicting PV power. Aprilla et al. [13] used convolutional neural network optimized by salp swarm algorithm (SSA-CNN) to forecast short-term PV power. Investigation into the different results between support vector machine optimized using salp swarm algorithm, long short-term memory optimized by salp swarm algorithm, and their proposed SSA-CNN model was done. The capabilities of their proposed model to adapt to the different weather conditions (heavy cloudy, light cloudy, rainy, and sunny days) were encouraging. In Niu et al. [14], random forest technique was used to select variables that are important in PV power forecasting. They then used improved grey ideal value approximation for similar day screening for the quality of data. The data is decomposed using complementary ensemble mode decomposition to stabilize the data before forecasting using optimized backpropagation neural network. The model performed better, but there are request for further studies concerning cloudy weather and rainy weather after improving on the model.

A study by Zhou et al. [15] used a hybrid long short-term memory and attention mechanism (LSTM-AM) to forecast PV power using time series data. Feature selection method of attention mechanism was applied to select variables relevant to PV power forecasting with the aim of enhancing the performance of the developed prediction model. Multilayer perceptron neural network, traditional long short-term memory, and autoregressive integrated moving average model combined with exogenous variable were compared to investigate the superiority of the proposed LSTM-AM model. Cervine et al. [16] tested a new method based on ANN and analog ensemble using data from Luxemburg and many synthetic solar stations. The hybrid model provided the best probabilistic result compared with individual ANN and analog ensemble methods. Shorter duration was one of the advantages of their proposed model. Bugala et al. [17] employed a new ANN network designed model called RBF 6: 6-5-1 for forecasting PV electricity generation. A study done by Wang et al. [18] proposed a model for forecasting PV power using environmental factors as input in the optimized support vector machine. Comparing the model with traditional support vector machine shows the accuracy level of the model developed. It is therefore evident in literature that there is scarcely use of dimensionality reduction techniques for selection of features for PV power forecasting.

This chapter systematically combines a supervised dimensionality reduction method called Neighborhood Component Analysis and optimized ANN with Particle Swarm Optimization (NCA-PSO-ANN). The motivation for creating such a hybrid NCA-PSO-ANN model includes the following:

1. Considering the complexities and challenges in PV prediction, this book chapter explores the systematic combination of supervised dimensionality reduction method of NCA and optimized ANN with PSO.
2. The motivation behind such a hybrid is that the NCA, which is a supervised dimensionality reduction method, can adequately identify and select the relevant features that contribute significantly to PV power and therefore enhances the prediction performance of PSO-ANN.

The proposed hybrid NCA-PSO-ANN leverage on their respective individual strengths and weaknesses to complement each other, which further improves the model's prediction outputs.

The remaining aspect of the chapter is as follows: Sect. 3.2 is dedicated to the proposed techniques, Sect. 3.3 handles the problem formulation, Sect. 3.4 presents numerical study, and finally, Sect. 3.5 presents the conclusion and recommendations.

3.2 Proposed Techniques

This section presents the foundations of the methods that were used to build the proposed hybrid NCA-PSO-ANN approach. The section concludes with the overview of the hybrid method, by proposing the structure of the full algorithm. It also presents the method PCA, which is used to build other variants hybrid prediction models for comparison purpose.

3.2.1 *Neighborhood Component Analysis*

NCA is a dimensionality reduction technique based on supervised learning. The technique belongs to the family of nonparametric techniques that has the ability to select the useful features (variables) of a data set [19–24]. In the ANN model development phase, the NCA is usually applied as a preprocessing technique for feature selection with the aim of increasing the forecasting accuracy of the model. Let us assume a data set of total data points N and input and output variables a and b. If the sample is D, then

$$D = \left(a_i, b_i\right), \quad i = 1, 2, 3, .., N. \tag{3.1}$$

The possibility of a_j being selected from the sample D is higher when a_j is nearer to the reference point a.The distance between a_j and a can be expressed as a function:

$$f_{\mathrm{d}}\left(a_i,a_j\right)=\sum_{g=1}^{N} w_g^2\left|a_{ig}-a_{jg}\right|, \tag{3.2}$$

where w_g represents the weights of the feature under consideration. Taking

$$P\left(a_j\,|D\right)\propto k\left[f_d\left(a_i,a_j\right)\right], \tag{3.3}$$

k is the similarity function which becomes large when $f_{\mathrm{d}}(a_i, a_j)$ value is small. Using this fact, Eq. 3.3 should be 1 for all j. We can therefore rewrite Eq. 3.3 as follows:

$$P\left(a_j\,|D\right)=\frac{k\left[f_{\mathrm{d}}\left(a_i,a_j\right)\right]}{\sum_{j=1,j\neq i}^{N} k\left[f_{\mathrm{d}}\left(a_i,a_j\right)\right]}. \tag{3.4}$$

Now let us assume that the forecasted values by randomized regression model and the observed value are O_{b} and O_{a}, respectively. If the loss function is denoted as l_f, then the mean value of $l(O_{\mathrm{b}}, O_{\mathrm{a}})$ is written as follows:

$$l_i=E\left[l\left(O_{\mathrm{b}},O_{\mathrm{a}}\right)\middle|D^{-i}\right]=\sum_{j=1,j\neq i}^{N} P_{ij}\left(O_i,O_j\right). \tag{3.5}$$

To eliminate overfitting of the model, a regularization term λ is introduced to the objective function. This makes the final objective function to be as follows:

$$f\left(h\right)=\frac{1}{N}\sum_{i=1}^{N} l_i+\lambda\sum_{g=1}^{p} w_g^2. \tag{3.6}$$

The pseudo-code for the NCA algorithm is presented in Table 3.1.

3.2.2 *Principal Component Analysis*

PCA is a method used for reducing the dimensionality of a data set. It is a known unsupervised technique where no information is needed before analyzing the data to reduce its dimension. PCA is good in keeping the issue of information loss to a minimum [25, 26] by considering all relevant variables that contribute to the main data and separating the irrelevant variables from the data set. PCA is known to possess a powerful advantage of keeping the trend and patterns embedded within the

Table 3.1 NCA algorithm

Set T: training set, α: initial step length, σ kernel width, λ: regularization parameter, η: positive constant
Initialize $\in^0 = -\infty$, $t = 0$, and $W^0 = (1,1,1,..,1)$
Repeat
For $i = 1,\ldots,N$ **do**
Calculate p_{ij} **and** P_i using W^t considering equations 3 and 4
For $l = 1,\ldots,d$ **do**
Calculate $\Delta l = 2\left[\frac{1}{\sigma}\sum_i (P_i \sum_{j \neq i} P_{ij} \lvert a_{il} - a_{jl} \rvert - \sum_j O_{ij} P_{ij} \lvert a_{il} - a_{jl} \rvert) - \lambda\right] w_l^t$
$t = t + 1$
$W^t = W^{t-1} + \Delta\alpha$
$\in^t = \xi W^{t-1}$
If $\in^t \rangle \in^{t-1}$ **then**
$\alpha = 1.01\alpha$ **else**
$\alpha = 0.4\alpha$
Until $\lvert \in^t - \in^{t-1} \rvert \langle \eta$
$W = W^t$
Return W
End

data set while reducing its dimensionality. After the dimensionality reduction process is concluded, the data becomes more simplified to analyze and visualize [27].

Let us assume that, F is the number of features that relates to PV power and is represented as $x_1, x_2, x_3, \ldots, x_f$. Then, applying the PCA, a linear pairing of the features $x_1, x_2, x_3, \ldots, x_f$ is performed to arrive at components n represented by $c_1, c_2, c_3, \ldots, c_n$. The principal component n contains most of the information from the base indexes. Equation 3.7 indicates the n components:

$$
\begin{cases}
C_1 = l_{11}x_1 + l_{12}x_2 + l_{13}x_3 + \ldots + l_{1f}xf \\
C_2 = l_{21}x_1 + l_{22}x_2 + l_{23}x_3 + \ldots + l_{2f}xf \\
C_3 = l_{31}x_1 + l_{32}x_2 + l_{33}x_3 + \ldots + l_{3f}xf \\
\vdots \\
C_n = l_{n1}x_1 + l_{n2}x_2 + l_{n3}x_3 + \ldots + l_{nf}xf
\end{cases}
\tag{3.7}
$$

From here, the indicator possessing the information that is the largest amount is assigned the principal component c_1. The second indicator is chosen if the first does not possess all the information associated with the original and labeled principal component c_2. It is important to note that c_2 contains different information from c_1. Therefore, covariance of the two is equal to zero and the variance of c_2 is smaller than c_1. This selection procedure is repeated until most of the base indicators information are retrieved [28–30].

3.2.3 *Particle Swarm Optimization*

PSO is a progressive meta-heuristic method that is suitable for continuous and discontinuous optimization problems. The PSO algorithm stems from school of fish swimming or flock of birds behavior looking for food together. It is a procedure to find possible solutions that a member of the population can represent at a time. Cooperation is a highly recognized factor in reaching the necessary level of intelligence possessed by the population [31–35]. A member is first chosen, and a location is allocated. This makes it possible to link every location to the optimization solution of the problem using each member's location. In view of the prior position and speed vector, every member position is always updated.

Considering PSO processes, let us describe x as the coordinate of a particle and its speed vector as v, while Gbest and Pbest are the best location between all members of the population and the best location settled by a member of the population respectively. All participating members are repeatedly updated on their new location until the optimum location is obtained. Equations 3.8 and 3.9 are used for updating rules of the particles velocity and position [36–39]:

$$V_i(t) = u_1 c_1 \left(x_{\text{Pbest}_i} - x_i(t) \right) + u_2 c_2 \left(x_{\text{Gbest}_i} - x_i(t) \right) + w v_i(t-1), \tag{3.8}$$

$$x_i(t) = v_i(t) + x_i(t-1), \tag{3.9}$$

where c is the coefficient of acceleration, t is time, w is inertia weight, and u is a uniformly distributed number whose value is between 0 and 1.

The inertia weight (w) can also be called inertia coefficient and is responsible for getting the optimal solution during the iterations by providing stability between global and personal explorations. The inertia weight is calculated by Eq. 3.10.

$$w = w_{\text{mx}} - \left[\frac{w_{\text{mx}} - w_{\text{mn}}}{\text{It}_{\text{mx}}} \right] \times \text{It}, \tag{3.10}$$

where w_{mx} is initial inertia weight, w_{mn} is final inertial weight, It_{mx} and It are the maximum and the current number of iterations. Figure 3.1 is the simplified PSO processes of finding the optimal position. Steps for the PSO processes are presented as follows:

1. The initial parameters such as population size, termination criteria, maximum velocity, and inertia weight required for the algorithm are determined.
2. Random population, which is equivalent to the specified population size at step 1, is generated. Every single member of the population must possess all the set parameters.
3. Then, the objective function for all the members of the population is set. The Pbest is the best solution achieved between the population during initial iteration.

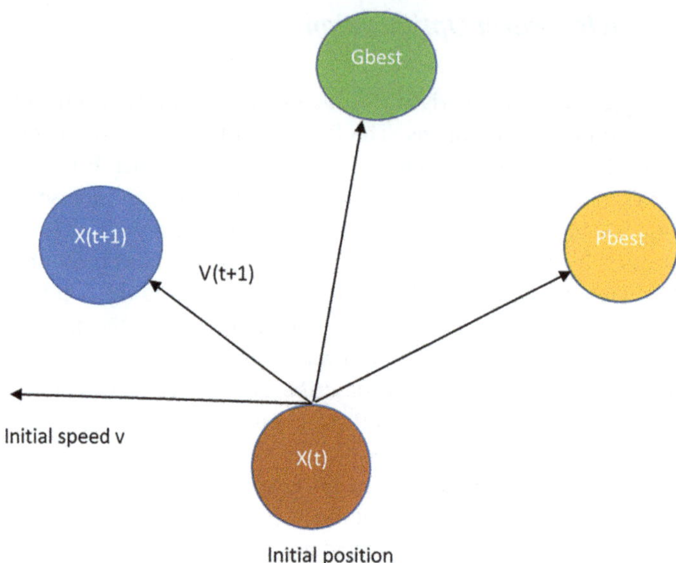

Fig. 3.1 Basic PSO structure

4. Each member's speed is updated and matched with the specified maximum speed. The speed must not exceed the maximum speed already set.
5. Position of each member is updated, and verify with upper and lower limits already set.
6. Each member's solution is matched, and the best for that member replaces Pbest. The best among the population becomes the Gbest or the optimal solution.
7. Stop when termination criteria are reached or else repeat from step 4.

3.2.4 Artificial Neural Network

ANN has proven to be a good method for solving prediction and classification problems [40]. In this chapter, we employed multilayer perceptron (MLP) neural network type of ANN. Its structure is feed-forward, which is made up of input layer, one or more hidden layer(s), and an output layer [41, 42]. It possesses the advantage of being universal approximation, which makes it more flexible and easily adaptable to nonlinear and complex tasks. The number of nodes associated with the input layer depends on the features involved, which contribute to the final prediction with each node of the connected layers assigned a weight [43, 44]. The weight is a vital factor during the learning process as the perceptron changes the assigned weight by comparing errors after each process until an acceptable error is achieved. The back-propagation algorithm is used for carrying out the weight changing task by the

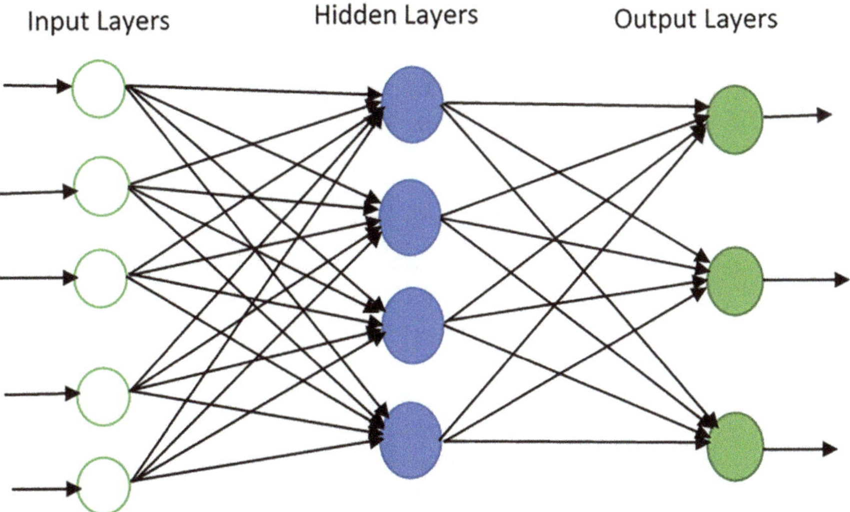

Fig. 3.2 Basic structure of MLP

perceptron [45, 46]. Basic structure of MLP is shown in Fig. 3.2. The MLP process can be mathematically expressed as follows:

$$y = \psi\left[\sum_{i=1}^{n} w_i x_i + B\right] = \psi\left[w^T x + B\right], \tag{3.11}$$

where ψ is the activation function, B is the bias, w is the vector weights, and x is the input vectors.

3.3 Problem Formulation

This section explains the hybrid models explored in this study for the purpose of forecasting PV power. To identify the most influential features on PV power forecasting, NCA is used to help select the most relevant features. The features that were rejected by the NCA were not used as input features for PV power forecasting. Another feature selection model that has been used in other studies is the PCA. In this chapter, PCA was explored to compare its performance to the NCA. Both NCA and PCA were carried out using MATLAB program. Figure 3.3 shows the entire process of the proposed hybrid NCA-PSO-ANN right from historical data harvesting to the end where the forecasting results were statistically analyzed.

To evaluate the prediction strength of the various models comprehensively, three performance indexes were used. These are the Mean Absolute Error (MAE), Percent Mean Absolute Relative Error (PMARE), and Root Mean Square Error (RMSE).

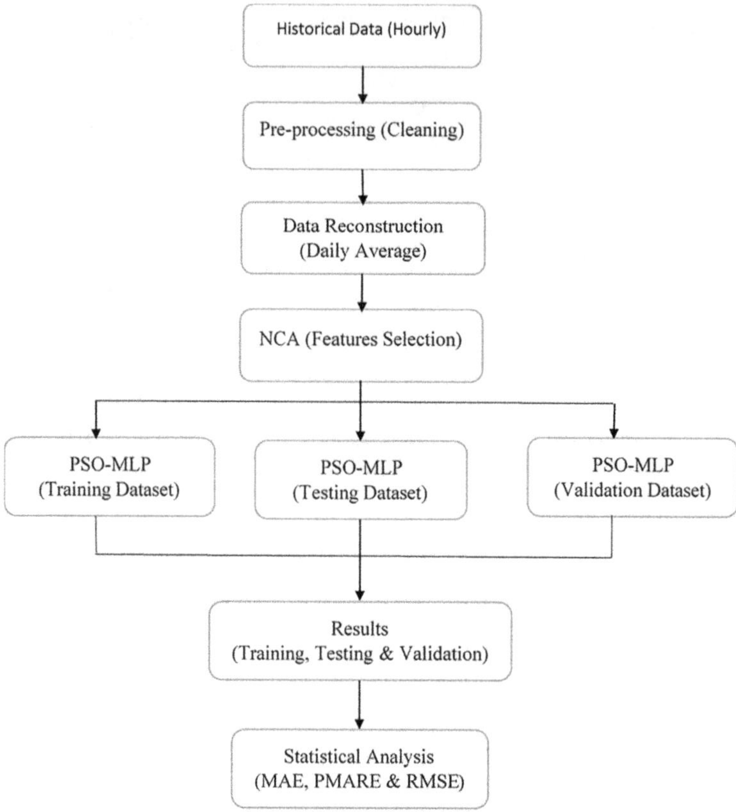

Fig. 3.3 Structure of the proposed NCA-PSO-ANN

The MAE was used to check the number of deviations present in the forecasted results. The PMARE was used to check the efficiency of the developed models, while RMSE was purposefully used to ascertain the dispersions in the forecasted results. Equations 3.12, 3.13, and 3.14 were, respectively, used for calculating MAE, PMARE, and RMSE.

$$\text{MAE} = \frac{\sum_{i=1}^{N}\left(A_i - F_i\right)}{N}, \tag{3.12}$$

$$\text{PMARE}(\%) = \frac{100}{N} \times \sum_{i=1}^{N} \frac{\left|A_i - F_i\right|}{A_i}, \tag{3.13}$$

$$\text{RMSE} = \sqrt{\frac{1}{N} \times \sum_{i=1}^{N}\left(A_i - F_i\right)^2}, \tag{3.14}$$

where N is the data set size, A is the actual data, and F is the forecasted data.

3.4 Numerical Study

3.4.1 *Characteristics of the Data Used for the Modeling*

For the purpose of checking the suitability of the proposed hybrid model (NCA-PSO-ANN), a real world data is sourced from Photovoltaic Geographical Information System (PVGIS) owned by European Commission Science Hub [47]. The base features of the data are as follows: longitude −2.489°, latitude 10.540°, elevation 303 m, optimal slope 14°, optimal Azimuth −10°, system power 1 kWp, and system losses 14%. The period during which the data was selected is 1 January 2014 to 31 December 2016. The data comprised six features including reflected irradiance on the inclined plane, sun height, direct beam irradiance on the inclined plane, diffused irradiance on the inclined plane, wind speed, and 2 m air temperature. Data cleaning was carried out to delete the hours when there is no PV power recorded because of nonavailability of the sun. The PV power is usually recorded during the hours of 07:07 am to 17:07 pm, which is when the sun rises and set during each day. The hourly data was converted to daily average data before the dimensionality reduction techniques of NCA and PCA were applied.

3.4.2 *Problem Simulation and Results Analysis*

The proposed NCA-PSO-ANN and the compared models of PCA-PSO-ANN, PSO-ANN, and PCA-ANN were simulated on MSI Prestige Series having the specifications: i7–8th Generation, 16.384 GB RAM, and 8 Central Processing Units (CPUs). A MATLAB 2019b program was used for the simulation process. The comparison is important to establish the performance capabilities of the various models. Moreover, the resultant effect of combining the dimensionality reduction techniques and the optimized ANN was ascertained.

The simulation results of the NCA revealed that four out of the six features considered were selected as the most relevant features that influence the PV power. Because NCA is a supervised dimensionality technique, it works by matching the input features to the output feature to determine the relevant input features that contribute to the success of predicting the output feature accurately. Figure 3.4 shows the features that were selected using NCA, where the features which have 0 feature weight are the rejected features. In Fig. 3.4, the selected features corresponding to numbers (1, 2, 5, and 6) were the direct beam irradiance on the inclined plane, diffused irradiance on the inclined plane, 2-m air temperature, and total wind speed at 10 m. The reflected irradiance on the inclined plane and the sun height were the rejected features. This implies that the two rejected features have minimal direct impact on PV power generation. Besides, the reflected irradiance is coming from the unabsorbed solar radiation that is reflected from the surface of the Earth. The sun height, on the other hand, is angular measurement of the sun relative to the

Earth's horizon and thus may not have so much effect on the PV power. The NCA selected features confirm to the normal practice by scholars involved in PV power modeling and prediction. Figure 3.5 shows the various regularization parameter (Lambda) NCA results. The role of this parameter is to prevent overfitting condition and produce an optimal Lambda value that can give the best feature selection. The optimal Lambda value in our work was 0.5 with the loss function (MSE) of 59 (Fig. 3.5). The MSE is the fitness function of the NCA objective function to determine the best Lambda value. In the model simulation, the NCA selected features become the input for the PSO-ANN and the output is the PV power.

For the PCA, only two principal components were selected. Figure 3.6 shows the various variations of the six principal components and their corresponding eigenvalues produced by the PCA. The eigenvalue is one criterion, which is also known as the Kaiser criterion, and was used to select the best principal components. It can be seen in Fig. 3.6 that principal components 1 and 2 achieved eigenvalues greater than one and thus were selected as the inputs into the ANN model building process. A further analysis of Fig. 3.6 indicates that the two selected principal components combined account for approximately 64.78% of the total variation of the six original input variables. That is, the two principal components served as the input data to the PSO-ANN, while the PV power is the target. It must be noted that in both NCA and PCA cases, the PSO was applied to optimize the training parameters of the ANN model to achieve the final prediction result.

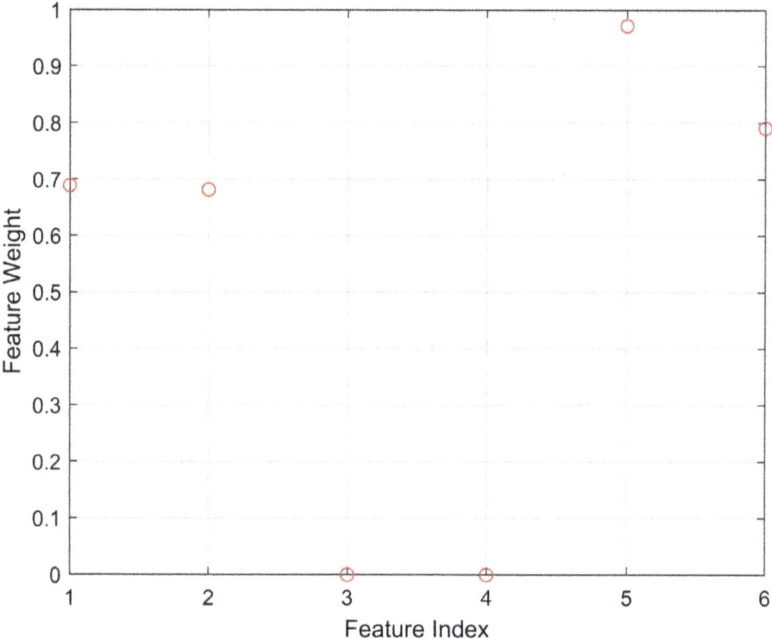

Fig. 3.4 NCA selected features

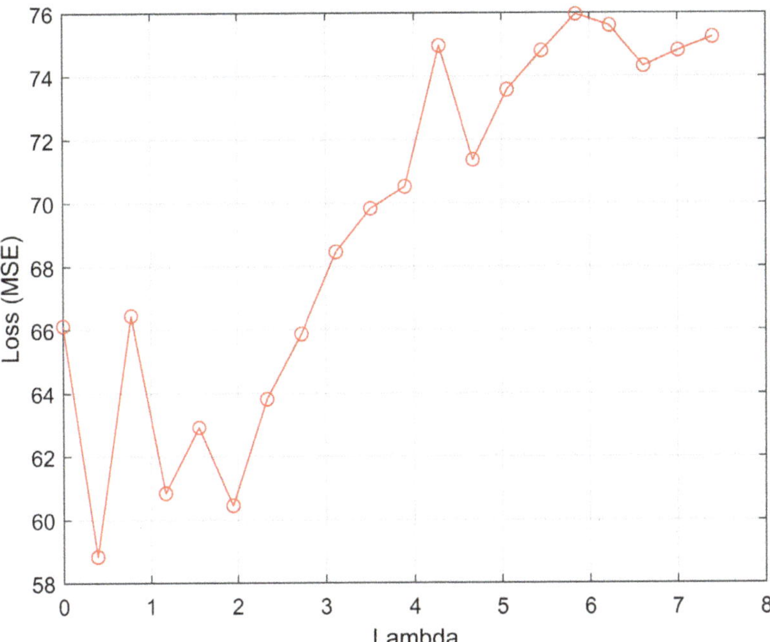

Fig. 3.5 NCA lambda result

The forecasted results from the proposed NCA-PSO-ANN model and three other models investigated (PCA-PSO-ANN, PCA-ANN, and PSO-ANN) are presented in Tables 3.2 and 3.3, respectively. The performance indicators used are also graphically represented in Figs. 3.7, 3.8, and 3.9 for the MAE, PMARE, and RMSE, respectively. The testing errors and validation errors are also shown in Figs. 3.10 and 3.11. The importance of utilizing both the dimensionality reduction and the optimized ANN have been proven by the results achieved in the study. The purpose of the validation results is to further investigate the prediction consistency of the proposed NCA-PSO-ANN model for handling the forecasting task considering the behavior of PV power. The validation errors are also important to the modeler to know that overfitting condition is avoided in the modeling process.

For the MAE results (Table 3.2), the proposed NCA-PSO-ANN model had the lowest score of 2.8808 in training, 3.2021 in testing, and 2.5489 in validation. The PCA-PSO-ANN model came second in training and validation but placed third in testing results with values of 5.7125 in training, 6.4639 in testing, and 7.0661 in validation. The PCA-ANN model followed with 3.7626 representing third position in training results, 6.3314 representing second position in testing, and 7.7685 that is third position in validation results. The PSO-ANN taking fourth position with 15.1944 in training, 17.1916 in testing, and 11.4354 in validation results. All these MAE results are graphically shown in Fig. 3.7, indicating superiority of the proposed NCA-PSO-ANN model which possesses minimum number of deviations in

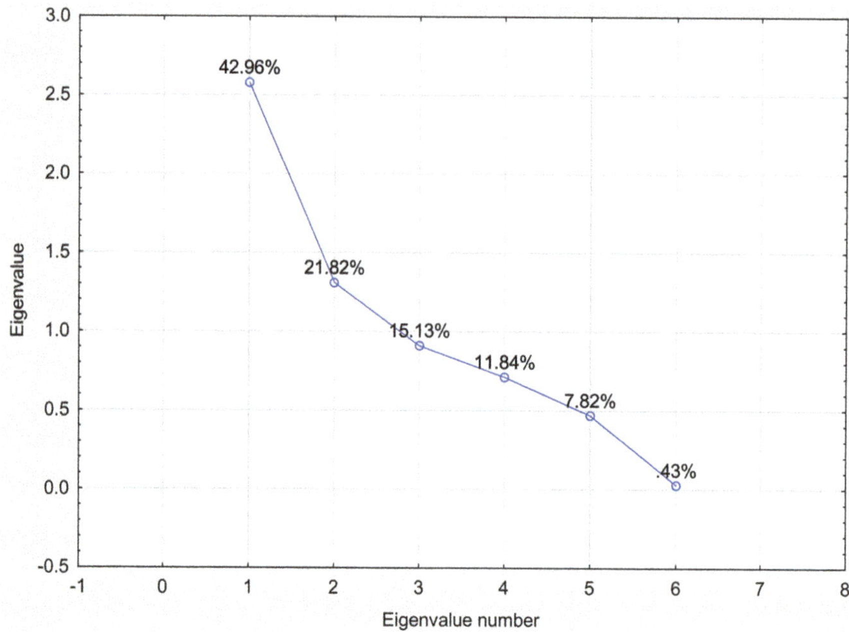

Fig. 3.6 A Screen plot of the amount of principal components variation

Table 3.2 PMARE and MAE results

	PMARE (%)			MAE		
MODEL	Training	Testing	Validation	Training	Testing	Validation
PCA-ANN	1.8752	2.0772	1.7567	6.7626	6.3314	7.7686
PSO-ANN	4.8869	6.1375	3.6809	15.1944	17.1916	11.4354
PCA-PSO-ANN	1.4741	2.0420	1.5635	5.7125	6.4639	7.0661
NCA-PSO-ANN	**0.9445**	**1.1785**	**0.5992**	**2.8808**	**3.3021**	**2.5489**

Table 3.3 RMSE results

	RMSE		
MODEL	Training	Testing	Validation
PCA-ANN	8.3518	8.6468	9.5464
PSO-ANN	19.7389	22.6524	19.9054
PCA-PSO-ANN	7.2022	8.9673	8.8527
NCA-PSO-ANN	**4.0379**	**4.2975**	**3.2311**

its results followed by PCA-PSO-ANN, PCA-ANN, and PSO-ANN in that sequence.

For the PMARE indicator, the proposed NCA-PSO-ANN model comparatively performed better than PCA-PSO-ANN, PCA-ANN, and PSO-ANN models. This is an indication that the NCA-PSO-ANN model is the most efficient than the other

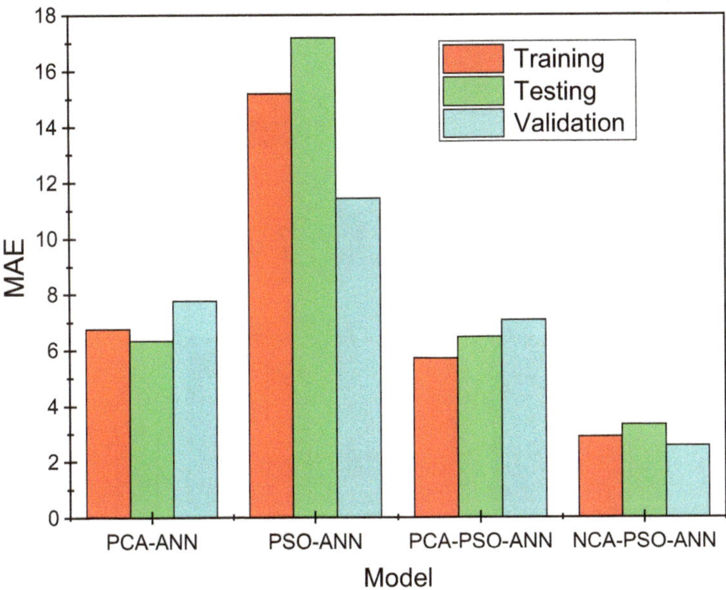

Fig. 3.7 MAE values obtained by the models

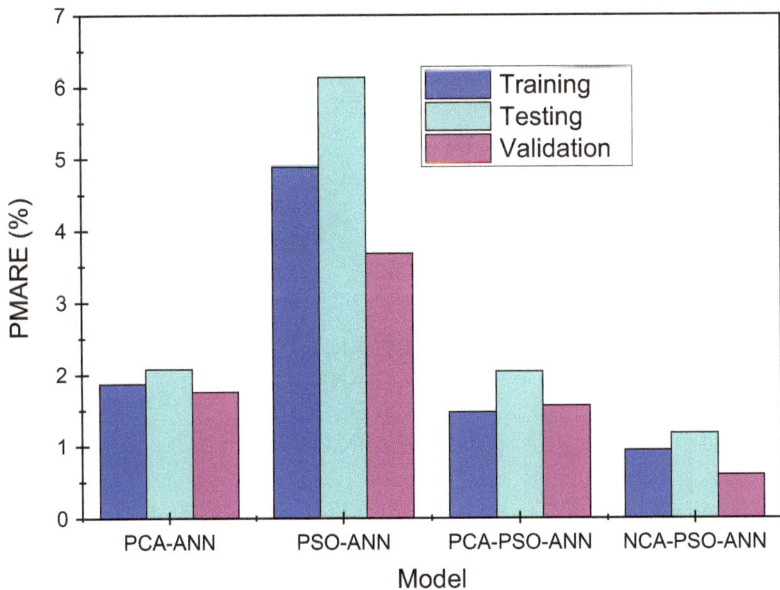

Fig. 3.8 PMARE values obtained by the models

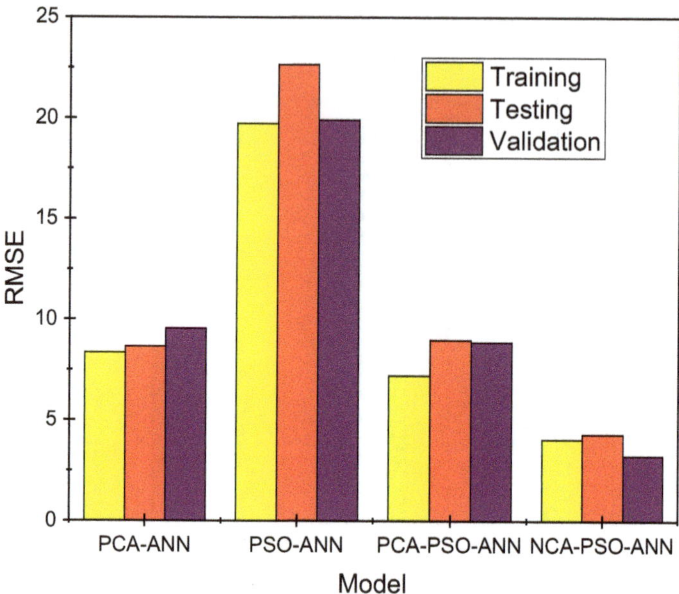

Fig. 3.9 RMSE values obtained by the models

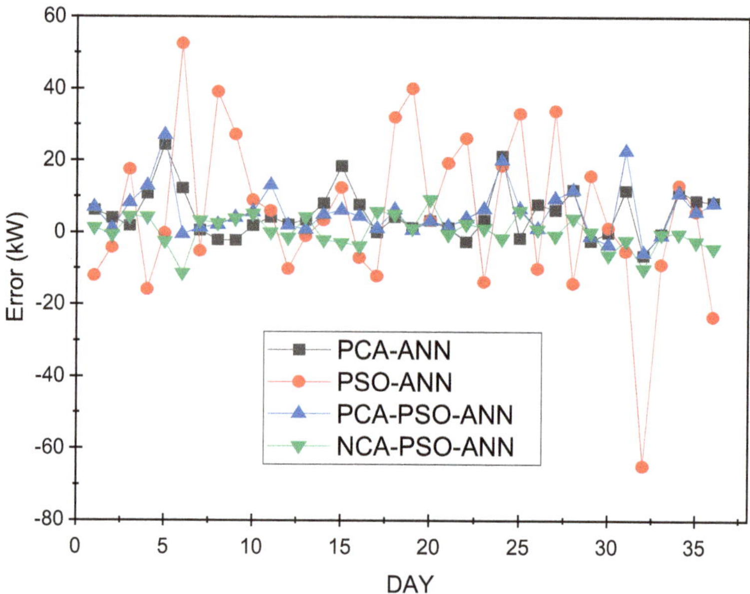

Fig. 3.10 Testing errors after forecasting using the models

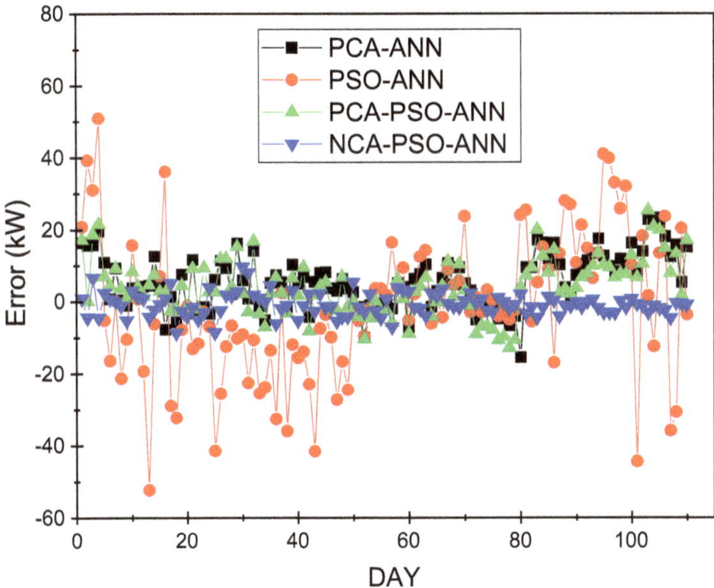

Fig. 3.11 Validation errors after forecasting

models. That is, the NCA-PSO-ANN produced PMARE values of 0.9445, 1.1785, and 0.5992% in training, testing, and validation, respectively. This means that the NCA-PSO-ANN model achieved 99.06, 98.82, and 99.40% efficiency in training, testing, and validation performance. This was followed by PCA-PSO-ANN model scoring efficiencies of 98.53, 97.96, and 98.44% for training, testing, and validation, respectively. The next model is the PCA-ANN, which had efficiencies of 98.12, 97.92, and 98.24%, while the PSO-ANN was the least performing model with efficiencies of 95.11, 93.86, and 96.32% corresponding to training, testing, and validation. Figure 3.8 graphically illustrates the PMARE results.

The RMSE indicator, which shows how disperse a forecasting result is from the actual values, was also used to check the prediction capabilities of the developed models. The proposed NCA-PSO-ANN model achieved the least values in all three categories, that is, training (4.0379 kW), testing (4.2975 kW), and validation (3.2311 kW). PCA-PSO-ANN trailed the NCA-PSO-ANN with RMSE values of 7.2022, 8.9673, and 8.8527 kW, respectively, for training, testing, and validation performance. The PCA-ANN once again outperformed PSO-ANN model with scores of 8.3518, 8.6468, and 9.5464 kW compared with PSO-MLP, which had 19.7389, 22.6524, and 19.9054 kW. The RMSE results are further presented for pictorial viewing in Fig. 3.9.

Figures 3.10 and 3.11 are the testing and validation errors, respectively, indicating how closed or dispersed the errors are from the ideal error value of zero (0). The NCA-PSO-ANN model's testing and validation errors (Figs. 3.10 and 3.11) are more closed to the ideal zero (0) error point on the graphs, while the other models'

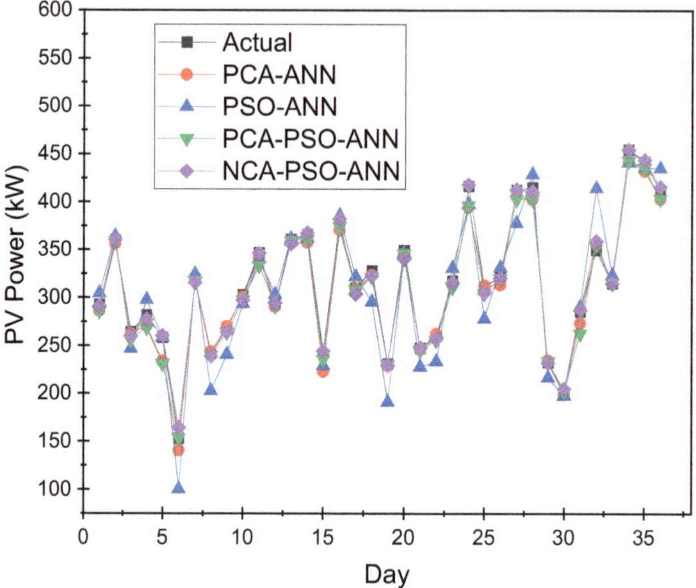

Fig. 3.12 Comparing the actual PV power with the forecasted PV power from the models

errors are more dispersed. From Fig. 3.12, comparing the forecasted results with the actual data, the NCA-PSO-ANN model demonstrated superiority with its predicted outcomes, conforming to the actual data pattern better than the remaining models investigated.

3.5 Conclusion and Recommendations

Considering the complex and nonstable characteristics of PV power, a novel approach by combining a supervised dimensionality technique of NCA and ANN optimized by PSO to form NCA-PSO-ANN was developed for forecasting PV power. The results indicate improvement in the forecasting performance by outperforming three other hybrid models. Considering the training, testing, and validation results, the following conclusions can be made:

1. The proposed NCA-PSO-ANN model was more efficient and reliable in handling the complexity and nonstationary characteristic of PV power data as compared to the three other hybrid models (PCA-ANN, PSO-ANN, and PCA-PSO-ANN) tested.
2. The combined effect of the feature selection and the optimization stages improved the performance accuracy of the NCA-PSO-ANN prediction model, indicating the usefulness of the proposed modelling strategy used in developing the model.

3. The complexity and nonstable characteristics of PV power are highly reduced for easy forecasting using the NCA and PSO for selecting features and optimizing the process.

Considering the forecasting results of the proposed model, it is recommended for policy makers, plant planners, and grid managers to utilize this model for more accurate information during forecasting PV power. Further studies that involve testing the capability of other optimization techniques to improve the accuracy of the ANN technique are recommended. Long-term forecasting of PV power using the proposed model is something that is being considered in the future to check its capabilities in that regard.

References

1. M. Rahmani-Andebili, *Cooperative Distributed Energy Scheduling in Microgrids.*, no. 9789811070006 (Springer Singapore, 2018)
2. Ö. Çelik, A. Tan, M. Inci, A. Teke, Improvement of energy harvesting capability in grid-connected photovoltaic micro-inverters. Energ. Sources, Part A Recover. Util. Environ. Eff. 1–25 (2020)
3. M. Morjaria, D. Anichkov, V. Chadliev, S. Soni, A grid-friendly plant: The role of utility-scale photovoltaic plants in grid stability and reliability. IEEE Power Energ. Mag. **12**(3), 87–95 (2014)
4. D. Manz, R. Walling, N. Miller, B. LaRose, R. D'Aquila, B. Daryanian, The grid of the future: Ten trends that will shape the grid over the next decade. IEEE Power Energ. Mag. **12**(3), 26–36 (2014)
5. F. Barbieri, S. Rajakaruna, A. Ghosh, Very short-term photovoltaic power forecasting with cloud modeling: A review. Renew. Sust. Energ. Rev. **75**(August 2015), 242–263 (2017)
6. J. Qu, Z. Qian, Y. Pei, Day-ahead hourly photovoltaic power forecasting using attention-based CNN-LSTM neural network embedded with multiple relevant and target variables prediction pattern. Energy **232**, 120996 (2021)
7. D.W. van der Meer, J. Widén, J. Munkhammar, Review on probabilistic forecasting of photovoltaic power production and electricity consumption. Renew. Sust. Energ. Rev. **81**, 1484–1512 (2018)
8. M. Rahmani-Andebili, Dynamic and adaptive reconfiguration of electrical distribution system including renewables applying stochastic model predictive control. IET Gener. Transm. Distrib. **11**(16), 3912–3921 (2017)
9. A. Mellit, A.M. Pavan, V. Lughi, Deep learning neural networks for short-term photovoltaic power forecasting. Renew. Energy **172**, 276–288 (2021)
10. M. Abdel-Basset, H. Hawash, R.K. Chakrabortty, M. Ryan, PV-Net: An innovative deep learning approach for efficient forecasting of short-term photovoltaic energy production. J. Clean. Prod. **303**, 127037 (2021)
11. Z.F. Liu, S.F. Luo, M.L. Tseng, H.M. Liu, L. Li, A.H.M. Mashud, Short-term photovoltaic power prediction on modal reconstruction: A novel hybrid model approach. Sustain. Energy Technol. Assess. **45**(October), 2020 (2021)
12. M.S. Hossain, H. Mahmood, Short-term photovoltaic power forecasting using an LSTM neural network and synthetic weather forecast. IEEE Access **8**, 172524–172533 (2020)
13. H. Aprillia, H.T. Yang, C.M. Huang, Short-term photovoltaic power forecasting using a convolutional neural network-salp swarm algorithm. Energies **13**(8) 1879 (2020)

14. D. Niu, K. Wang, L. Sun, J. Wu, X. Xu, Short-term photovoltaic power generation forecasting based on random forest feature selection and CEEMD: A case study. Appl. Soft Comput. **93**, 106389 (2020)
15. H. Zhou, Y. Zhang, L. Yang, Q. Liu, K. Yan, Y. Du, Short-term photovoltaic power forecasting based on long short term memory neural network and attention mechanism. IEEE Access **7**, 78063–78074 (2019)
16. G. Cervone, L. Clemente-Harding, S. Alessandrini, L. Delle Monache, Short-term photovoltaic power forecasting using artificial neural networks and an analog ensemble. Renew. Energy **108**, 274–286 (2017)
17. A. Bugała et al., Short-term forecast of generation of electric energy in photovoltaic systems. Renew. Sust. Energ. Rev. **81**, 306–312 (2018)
18. J. Wang, R. Ran, Z. Song, J. Sun, Short-term photovoltaic power generation forecasting based on environmental factors and GA-SVM. J. Electr. Eng. Technol. **12**(1), 64–71 (2017)
19. O. Yaman, An automated faults classification method based on binary pattern and neighborhood component analysis using induction motor. Meas. J. Int. Meas. Confed. **168**(August), 2020 (2020)
20. H. Zhou, J. Chen, G. Dong, H. Wang, H. Yuan, Bearing fault recognition method based on neighbourhood component analysis and coupled hidden Markov model. Mech. Syst. Signal Process. **66–67**, 568–581 (2015)
21. S. Raghu, N. Sriraam, Classification of focal and non-focal EEG signals using neighborhood component analysis and machine learning algorithms. Expert Syst. Appl. **113**, 18–32 (2018)
22. N.S. Malan, S. Sharma, Feature selection using regularized neighbourhood component analysis to enhance the classification performance of motor imagery signals. Comput. Biol. Med. **107**(October 2018), 118–126 (2019)
23. W. Yang, K. Wang, W. Zuo, Neighborhood component feature selection for high-dimensional data. J. Comput. **7**(1), 162–168 (2012)
24. T. Tuncer, F. Ertam, Neighborhood component analysis and reliefF based survival recognition methods for hepatocellular carcinoma. Phys. A Stat. Mech. Appl. **540,** 123143 (2020)
25. F. Cui, M. Kim, C. Park, D. Kim, K. Mo, M. Kim, Application of principal component analysis (PCA) to the assessment of parameter correlations in the partial-nitrification process using aerobic granular sludge. J. Environ. Manag. **288,** 112408 (January) (2021)
26. S. Li, T. Chen, L. Wang, C. Ming, Effective tourist volume forecasting supported by PCA and improved BPNN using Baidu index. Tour. Manag. **68,** 116–126 (2018)
27. E. Odhiambo Omuya, G. Onyango Okeyo, M. Waema Kimwele, Feature selection for classification using principal component analysis and information gain. Expert Syst. Appl. **174,** 114765 (January) (2021)
28. M. He, Y. Zhang, D. Wen, Y. Wang, Forecasting crude oil prices: A scaled PCA approach. Energy Econ. **97,** 105189 (2021)
29. L. Zhang, J. Wang, Q. Duan, Estimation for fish mass using image analysis and neural network. Comput. Electron. Agric. **173,** 105439 (April) (2020)
30. X. Wen, Z. Xu, Wind turbine fault diagnosis based on ReliefF-PCA and DNN. Expert Syst. Appl. **178,** 115016 (March) (2021)
31. H. Basser et al., Hybrid ANFIS-PSO approach for predicting optimum parameters of a protective spur dike. Appl. Soft Comput. J. **30**, 642–649 (2015)
32. H.M.I. Pousinho, V.M.F. Mendes, J.P.S. Catalão, A hybrid pso-anfis approach for short-term wind power prediction in Portugal. Energy Convers. Manag. **52**(1), 397–402 (2011)
33. L. Zhang, L. Zhao, High-quality face image generation using particle swarm optimization-based generative adversarial networks. Futur. Gener. Comput. Syst. **122**, 98–104 (2021)
34. S. Li, Q. Zhang, Z. Zhang, Q. Zhao, L. Liang, Improved subgroup method coupled with particle swarm optimization algorithm for intra-pellet non-uniform temperature distribution problem. Ann. Nucl. Energy **153,** 108070 (2021)
35. M. Jafari, E. Salajegheh, J. Salajegheh, Optimal design of truss structures using a hybrid method based on particle swarm optimizer and cultural algorithm. Structure **32**, 391–405 (2021)

36. M.E. Barrios Aguilar, D.V. Coury, R. Reginatto, R.M. Monaro, Multi-objective PSO applied to PI control of DFIG wind turbine under electrical fault conditions. Electr. Power Syst. Res. **180,** 106081 (2020)
37. H.S. Pannu, D. Singh, A.K. Malhi, Multi-objective particle swarm optimization-based adaptive neuro-fuzzy inference system for benzene monitoring. Neural Comput. Appl. **31**(7), 2195–2205 (2019)
38. E. Ofori-Ntow Jnr, Y.Y. Ziggah, S. Relvas, Hybrid ensemble intelligent model based on wavelet transform, swarm intelligence and artificial neural network for electricity demand forecasting. Sustain. Cities Soc. **66,** 102679 (2021)
39. J. Xu, W. Tan, T. Li, Predicting fan blade icing by using particle swarm optimization and support vector machine algorithm. Comput. Electr. Eng. **87,** 106751 (2020)
40. B. Singh, Predicting airline passengers' loyalty using artificial neural network theory. J. Air Transp. Manag. **94,** 102080 (April 2020) (2021)
41. U. Orhan, M. Hekim, M. Ozer, EEG signals classification using the K-means clustering and a multilayer perceptron neural network model. Expert Syst. Appl. **38**(10), 13475–13481 (2011)
42. A. Hashemi Fath, F. Madanifar, M. Abbasi, Implementation of multilayer perceptron (MLP) and radial basis function (RBF) neural networks to predict solution gas-oil ratio of crude oil systems. Petroleum **6**(1), 80–91 (2020)
43. L.C.P. Velasco, R.P. Serquiña, M.S.A. Abdul Zamad, B.F. Juanico, J.C. Lomocso, Week-ahead rainfall forecasting using multilayer perceptron neural network. Procedia Comput. Sci. **161,** 386–397 (2019)
44. A.A. Ewees, M.A. Elaziz, Z. Alameer, H. Ye, Z. Jianhua, Improving multilayer perceptron neural network using chaotic grasshopper optimization algorithm to forecast iron ore price volatility. Resour. Policy **65,** 101555 (February 2019) (2020)
45. A. Joshuva, R.S. Kumar, S. Sivakumar, G. Deenadayalan, R. Vishnuvardhan, An insight on VMD for diagnosing wind turbine blade faults using C4.5 as feature selection and discriminating through multilayer perceptron. Alexandria Eng. J. **59**(5), 3863–3879 (2020)
46. A.A. Heidari, H. Faris, I. Aljarah, S. Mirjalili, An efficient hybrid multilayer perceptron neural network with grasshopper optimization. Soft. Comput. **23**(17), 7941–7958 (2019)
47. T. Huld, R. Müller, A. Gambardella, A new solar radiation database for estimating PV performance in Europe and Africa. Sol. Energy **86**(6), 1803–1815 (2012)

Chapter 4
Applications of Artificial Intelligence in Short-Term and Long-Term Forecasting Techniques

Serkan Ayvaz

Abstract Electrical energy, due to immediate consumption requirements, is a type of energy that needs to be planned, produced, and transmitted in a quick and efficient way. Short-term and long-term electricity forecasts are vital for energy providers, regulators, and consumers when planning and operating grids, estimating tariffs, and calculating energy supply and demand. This chapter explores the applications of artificial intelligence in short-term and long-term forecasting techniques including, but not limited to, statistical methods to deep learning algorithms. Various artificial intelligence algorithms and methods used in the applications in the spectrum of time-dependent electricity consumption to smart grid planning and operations are investigated extensively. Moreover, the properties of time series data, fundamental concepts of machine learning, and techniques of hyperparameter tuning are described in detail.

Keywords Short-Term Forecasting (STF) · Long-Term Forecasting (LTF) · Smart Grid · Artificial intelligence (AI) · Big Data · Neural networks (NN)

4.1 Introduction

Electrical energy is a type of energy that must be transmitted efficiently and quickly due to its need for immediate consumption. The demand for electricity is rapidly growing as a result of multiple factors including urbanization, global population

S. Ayvaz (✉)
Department of Computer Engineering, Yildiz Technical University, Istanbul, Turkey

Department of Artificial Intelligence Engineering, Bahcesehir University, Istanbul, Turkey
e-mail: sayvaz@yildiz.edu.tr

M. Rahmani-Andebili (ed.), *Applications of Artificial Intelligence in Planning and Operation of Smart Grids*, Power Systems,
https://doi.org/10.1007/978-3-030-94522-0_4

growth, industrialization of countries, advances in technology, and increasing usage of technological devices in daily life [1]. Consequently, the forecasting of electricity accurately and effectively has become even more important for the management of electricity grids [2], planning of power generation [3], distribution [4], and transmission of electricity to meet the load and demand [5].

Continuous increase in electrical energy demand, limited resources, and the inability to store electricity efficiently compels the stakeholders to make better decisions and plans. It is necessary to maintain the production/consumption balance during production, as it not only provides significant financial gains but also causes surplus of electricity production, and overconsumption of power can lead to serious problems [6].

In recent years, forecasting electricity production and demand has become a pressing topic, especially for countries with inadequate resources to respond to soaring consumption [7]. Moreover, environmental issues, including global warming and international factors such as dependency on foreign energy sources, raise the level of awareness and motivation for saving electricity and preventing excessive energy production.

To help slow the climate change, it is essential to lower the global greenhouse gas emissions in electricity generation by increasing the use of renewable energy sources [8]. Smart Grids are key to integrate electricity generated by multiple renewable energy sources into a multimodal power system [9]. However, the power produced from renewable energy sources such as wind, hydroelectric, and solar tend to have a great deal of variability [4, 10]. It is necessary to reduce the uncertainty of power output from the renewable energy sources to make it practicable for operational tasks of estimating load demand, scheduling in grids [2]. Various forecasting and optimization approaches have been proposed to overcome the problem of variability in power output [3, 4, 8, 11].

Under these conditions, Information and Communication Technologies (ICTs) are at the heart of the fields of management and planning of supply systems, regulations of tariffs, operations of grids, and monitoring demand load of consumers [12]. Various ICT-based estimation methods were developed to predict the electricity production, consumption, storage, and distribution to market [4, 13].

Significant changes have occurred recently such as developments in Internet-of-Things (IoT) and renewable energy technologies, reducing costs of sensors and smart devices, and ease of access to these devices. These changes have led to an increase in the amount and diversity of connected devices, including smart meters, sensors, and smart devices that are built into a growing number of buildings and factories [14]. Electric vehicles can also be considered as smart devices due to their constant need of charging. Consequently, the electricity grids are evolving to smart grids. All of these devices are producing massive amount of data. For effective planning, operating, and maintaining such systems, device generated big data must be analyzed, which involves processing data through data mining and artificial intelligence techniques. This is a challenging task and even more so when performed in real time.

The objective of this book chapter is to provide in-depth coverage of the techniques and applications of short-term and long-term electricity forecasting. Thus, the needs and challenges in electricity forecasting are described in detail. The chapter aims to focus on the state-of-the-art forecasting algorithms and explore the recent key technological advances in electricity forecasting implementations. By reviewing the main challenges, important factors, useful applications investigated by studies, and different implementation strategies applied to the electricity forecasting systematically, this chapter intends to present the context for the readers and provide them the ground for further investigations for making informed decisions.

The main contributions of the chapter can be described as follows. The chapter investigated the underlying theoretical principles of the artificial intelligence methods used in short-term and long-term electricity forecasting and their applications to a great extent. Additionally, it explored the general characteristics of electricity data and the associated difficulties faced when analyzing the data in the form of a time series. Furthermore, the chapter implemented a use case based on a real-world dataset with the goal of demonstrating the importance of short-term power demand forecasting for an electricity distribution company. The performances of the forecasting algorithms were evaluated in a comparative approach as a practical application.

The remaining of the chapter is organized as follows. Section 4.2 reviews the requirements and challenges in electricity forecasting domain to introduce the context. In Sect. 4.3, the properties of electricity data and data preprocessing tasks performed in time series data are discussed in detail. In Sect. 4.4, various forecasting algorithms and implementations are systematically investigated. Additionally, evaluation metrics and hyperparameter tuning techniques are presented in a comparative way in Sect. 4.5. Lastly, the importance of the study is briefly summarized and finalized with concluding remarks.

4.2 Requirements and Challenges in Electricity Forecasting

Electricity forecasting can be investigated from various aspects. From the scale of the prediction unit, the forecasting can be performed at a meter level, building level, district level, or region level [14]. At each of these levels, different challenges are encountered during forecasting. There are numerous factors that have short-term and long-term effects, which make the predictability difficult [15].

With the recent advancements in IoT, smart grid and smart meter technologies, it has now become possible to make meter-level forecasts. Since smart grid technologies are relatively new and still emerging as a field of research, most forecasting applications and academic research have been focused on large-scale (i.e., district or regional level forecasting) rather than meter-level predictions [16].

Meter and building level forecasting is more challenging as the degree of variations is larger and thus more difficult to estimate with high accuracy [17, 18]. There are countless factors that can cause unexpected fluctuations [19]. To name a few: insulation type, architectural plan and location of the building unit, the number and usage habits of residents, and the type and amount of appliances can be considered among many other factors that significantly influence forecasting results [20].

Due to dynamic patterns, granularity, and high variations in meter and building scale electricity data, conventionally used statistical time series analysis and regression forecasting models are not always capable of capturing patterns and obtaining successful prediction results. Hence, research studies have been increasingly focusing on more advanced methods, for example, deep recurrent neural networks that have the capacity to learn patterns in long sequences [17, 18, 21–23].

The difficulties are not limited to the meter and building level forecasting in granular electricity data. District and region level large-scale forecasting applications are also affected by a large number of similar variables [15]. The major factors influencing electricity load forecasting can be broadly categorized as meteorological, seasonal, socioeconomical, consumer-related factors, and random factors [16].

4.2.1 *Classification of Electricity Forecast Intervals*

Although there is no standard classification of electricity forecasting periods, there is a general consensus on the types of forecast intervals [24–27]:

- Very short-term forecasting (VSTF): Forecasting a time period less than 24 hours. It generally uses only past loads.
- Short-term forecasting (STF): Forecasting a time period greater than 24 hours up to 1 week, using weather data with past loads.
- Medium-term forecasting (MTF): Forecasting a time period between 1 week up and 1 year, using economic information, weather data, and past loads.
- Long-term forecasting (LTF): Forecasting a time period longer than 1 year, using weather data, economic information, and demographic and occasionally spatial information with past data.

Short-term forecasting and very short-term forecasting fields have been explored more extensively when compared to long-term forecasting in literature; the same holds for review articles published in recent decades.

By nature, short-term and very short-term forecasting are typically used for applications performing daily operations. Thus, the predictions must be very accurate for these applications. On the other hand, medium- and long-term forecasting can tolerate some prediction inaccuracies in relative to short-term forecasting as they are usually applied in planning long-term objectives such as project planning system capacity, grid expansions, system maintenance, and cost estimations of operations [26].

4.3 Properties of Electricity Data and Data Preparation

4.3.1 Time Series Overview

Datasets associated with discrete-time values are usually defined as time series. Time series data possess the properties of sequences to the extent that the dynamic behavior of each time observation depends on the previous observations. The sequences also have pattern and meaning themselves. Performing predictions based on past time observations is called time series analysis.

In time series data, numeric attributes can be recorded as discrete or continuous types. In discrete forms, the observations are represented as enumerable values in time durations such as hourly, daily, weekly, monthly, or annual intervals. However, some observations are continuous, which can only be measured in approximation within a range of infinitely many values such as temperature, humidity, and power consumption.

4.3.1.1 Decomposition of Time Series: Seasonal Variations/Trend/Cyclic Variations/Irregularities

Time series data are known to have four different variations. These are, namely, trend, seasonality, and cyclic and irregular fluctuations. Trend component in time series is the long-term variations of the mean of the observations. In long term, the series may present an increasing or decreasing trend even when it may not be clear due to short-term fluctuations. However, the direction of trend is not necessarily linear over the series. It is possible that the trend may change from an increasing to a decreasing trend relative to the observations [28]. An increasing trend is illustrated using a sample dataset as shown in Fig. 4.1. The figure demonstrates the decomposition of time series components for illustrative purposes.

Time series data may have seasonal patterns repeating over fixed period of times. Seasonality can occur in short-term periods, for instance, daily or weekly variations, or long-term intervals such as the months of the year [29]. Seasonality can be clearly observed that repeats over fixed time periods as presented in Fig. 4.1 using a sample dataset.

Occasionally, cyclic patterns may occur in time series data. Although they are often confused with seasonality, cyclic component presents different characteristics. Unlike seasonality, cyclic variations do not take place with a known frequency during fixed time intervals. The cycles tend to take longer than the seasonal periods and usually present larger variations compared to the latter [30]. Yet, it is still possible to model the cyclic patterns for forecasting. Cyclic patterns can emerge due to external factors such as environmental, economic, and political conditions.

Even after excluding seasonal, trend, and cyclic variations, some irregular fluctuations may remain in time series data that can occur randomly as a consequence of unknown factors and are not predictable due to the nature variations [31] .

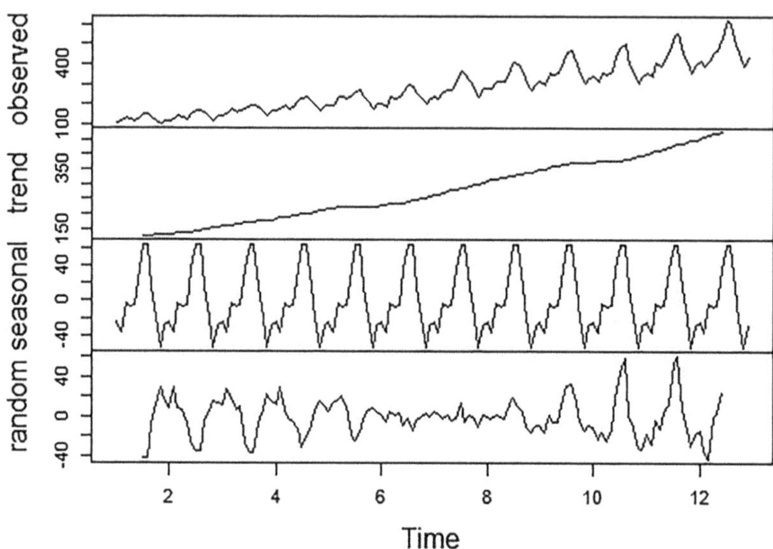

Fig. 4.1 A sample of time series decomposition

It is beneficial to observe the components of the time series data and try to understand how these components influence the patterns of the observations before proceeding to the selection of forecasting methods [28]. It can help with choosing suitable forecasting techniques that are capable of adequately modeling these characteristics.

4.3.2 Stationarity

Stationarity can be considered as one of the most crucial characteristics of time series data. Stationarity implies that statistical properties such as variance, mean, and autocorrelation of time series data do not change with time and, thus, they are in statistical equilibrium [32]. In other words, the time series can vary over time, but the rate of change is constant over time. Many statistical methods of time series analysis assume the data to be stationary. Stationary series do not contain trend or seasonality.

To determine the stationarity of time series data, statistical Unit root tests are typically applied. Augmented Dickey-Fuller (ADF) [33], Kwiatkowski-Phillips-Schmidt-Shin (KPSS) [32], Phillips–Perron [34], and Augmented Dickey-Fuller with Generalized Least Squares (ADF-GLS) [35] tests are among the most commonly used and effective Unit root tests. These statistical hypothesis tests are general purpose methods that help decide the rejection or validity of the null hypothesis. That being said, it is possible that the autocorrelation output may indicate seasonality in the time series, while the test results imply the dataset as stationary [36].

Different Unit root tests can have diverse assumptions, which are prone to result in conflicting interpretations [37]. Most of these tests (as in the case of ADF) assume that the time series is not stationary as their null hypothesis and attempt to test the evidence to reject the null hypothesis [36]. If small p-value, i.e., a value less than 0.05, is obtained, the results indicate that the null hypothesis is rejected and that the time series is stationary.

On the contrary, the null hypothesis of KPSS is the opposite. KPSS tests assume that the time series is stationary as its null hypothesis. The rejection of null hypothesis means that the dataset is nonstationary. More specifically, if large p-value, that is, a value greater than 0.05, is obtained then the null hypothesis of the test is accepted and the series is stationary [36].

Another method to determine stationarity of time series is observing the AutoCorrelation Function (ACF) plot, which is a visual representation of how autocorrelation progresses over the time series data [28]. In other terms, ACF plot demonstrates how the time series correlates with itself at different time points. In nonstationary, the ACF plot shows a gradual decrease of values and r_1 value is likely to have a large positive score. Conversely, in the case of stationary time series, the values on the ACF plot would suddenly drop to near zero. If the time series is not stationary, some statistical techniques such as curve fitting, differencing, and data transformation can be applied to make it stationary [38].

A common method to remove long term trend from time series is fitting a regression model, for example, linear line, to the data. Consequently, the residuals are retrieved from the model and used in place of the original points to separate the trend from the data [39].

Secondly, if the variance in the time series is not constant, variance-stabilizing transformations including logarithmic, square root, and multiplicative inverse functions can be applied to help reduce the variance in the data.

4.3.2.1 Differencing

Another widely used technique to make time series data stationary is differencing, which simply subtracts the values between consecutive observations in the sequence. Since the differencing method only considers the differences of values, it can stabilize the series by removing trend and seasonality, thereby reducing the variance in the data [28].

In first-order differencing, the new values, denoted by y'_t, are computed by subtracting current value from previous observation in the original data, denoted by y_t, to make the time series stationary as expressed in Eq. 4.1:

$$y'_t = y_t - y_{t-1} \tag{4.1}$$

More rarely, second-order differencing may be required to generate a stationary series. However, it is important to note that second-order differencing y''_t is not the subtraction of values from two previous time lags; instead, it is calculated as the change in the changes as demonstrated in Eq. 4.2:

$$\begin{aligned} y''_t &= y'_t - y'_{t-1} \\ &= (y_t - y_{t-1}) - (y_{t-1} - y_{t-2}) \\ &= y_t - 2y_{t-1} - y_{t-2} \end{aligned} \tag{4.2}$$

4.3.3 Preprocessing Electricity Data

In time-dependent electricity data, several data preprocessing tasks often needed to be performed before developing models. Performing data preparation tasks can be a tedious process, but proper preprocessing of data is of utmost importance to the success of machine learning models. Figure 4.2 presents a flowchart illustrating an overview of the steps followed in data preparation and model development. To clearly observe the seasonal and trend features of time series data, long periods of observations should be inspected. While using statistical time series analysis techniques such as ARIMA and its variants, data must be checked for stationarity and transformed to make it stationary, if necessary.

When having an abundance of features in multivariate data, feature selection methods can be performed. With the aim of reducing the number of input features to related variables, feature selection methods help improve the scalability, robustness, and prediction power of models. Lasso Regularization [40], Stepwise regression [41], Regularized Trees [42], and Principle Component Analysis [43] can be named among extensively utilized techniques.

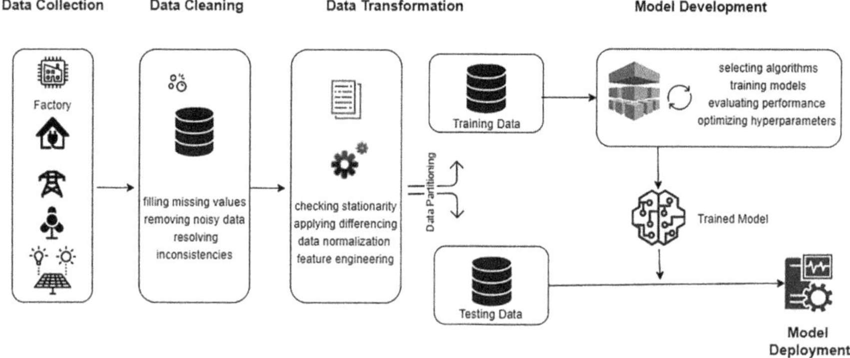

Fig. 4.2 A flowchart depicting the tasks in data preparation and modeling

Additionally, data normalization of time series with high variance is a critical process during modeling. Data normalization can be applied to necessary attributes by using techniques such as standardization and min-max normalization, which transform the values to their equivalents between 0 and 1. By converting variables in a dataset into the same scale, data normalization is instrumental in elimination of possible optimization issues.

The data resampling process in time series datasets is frequently used. It helps better understand the data during the analysis phase and thus obtain more accurate results. This process can be defined as configuration of a data set to a lower or higher frequency.

To examine potential overfitting and underfitting problems, a proper data splitting technique such as keeping validation dataset or applying cross-validation should be taken into account. While there is no universal strategy for this process, studies often split datasets into various ratios including splitting fifty/fifty, eighty/twenty, seventy/thirty, or applying k-fold cross validation.

Another preprocessing step to consider when modeling time series data using deep learning algorithms is applying various transformations to prepare the input data shape according to the modeling method. Particularly in RNN methods such as LSTM and GRU, the data set provided to the input layer must be represented as a 3D matrix consisting of time steps and number of attributes and the number of samples, respectively.

4.4 Methods of Short-Term and Long-Term Forecasting

Electricity forecasting has been extensively investigated by researchers from the research community. From different perspectives, numerous methods have been applied to electricity forecasting applications [3, 19, 20, 44, 45]. To electricity forecasting problem, a great deal of researchers approached it from a regression analysis perspective in high-dimensional data [46–50], while many others have considered it as a statistical time series forecasting issue [51–55], and more commonly as a supervised prediction task in machine learning [17, 21, 56–59].

In electricity forecasting, it is usually necessary to correctly estimate the future observations numerically with a small degree of error rather well, so that planning and other actions can be taken ahead of time [4]. Thus, the problem is typically considered as a regression problem [46–48, 50, 60].

On the other hand, majority of forecasting tasks center on the analysis of time series as the electricity data are recorded time-dependently and dynamic structure of data forms sequence. In other words, the goals of those models are to predict future observations in time-dependent data by propagating through time and minimizing potential forecast errors from previous input time-steps. This section provides a theoretical framework for the methods used in short-term and long-term electricity forecasting along with a review of applications in the context of the methods.

4.4.1 Statistical Methods

Short-term and long-term forecasting techniques have been utilized in the industry as common tools to plan production, sustain operational management, and reduce maintenance costs in grids. Numerous tasks such as energy production, distribution and maintenance in grids, and measurement of load consumption of energy require effective forecasting. A large number of research studies have proposed diverse theoretical and practical applications.

Although recent applications widely use different machine learning, deep learning models or their hybrid combinations, early forecasting applications focused more on statistical methods. Inherently, electricity data often represented in sequence and in form of time series. Frequently reported prediction methods in time series include Autoregressive and Moving-Average (ARMA) methods and their derivatives [51, 52]. Having said that, it is widely recognized that machine learning methods usually outperform the ARMA models in terms of prediction accuracy and that they are more flexible and generalizable than their ARMA counterparts [52, 61].

4.4.1.1 Regression Models

Regression analysis is a well-known statistical approach that can learn patterns from past observations and project them to a mathematical equation to predict values of a target variable in the future. Multiple linear regression (MLR) is a generalized form of regression that is represented as a linear equation and it can leverage multiple independent features when forecasting the dependent variable, which must take a numerical value [41]. The generalized form of an MLR model can be formulated as in the Eq. 4.3:

$$y_t = \alpha + \beta_1 x_1 + \beta_2 x_2 + \beta_3 x_3 + \ldots + \beta_n x_n, \tag{4.3}$$

where y_t denotes the target variable and x_i values are independent features of the model. Unknown parameters β_i represent the corresponding coefficients of the variables that are estimated during the model fitting.

The basic assumption and main limitation of MLR is the presence of a linear relationship between independent features and the target variable. In reality, many processes and factors interact in a nonlinear manner. Thus, in cases of nonlinearity of data, regression modeling with linearity assumption only approximates to real observations in the future and results in a large amount of error in accuracy.

Several types of tricks have been developed to provide more flexible models and better fit the data with nonlinear patterns: examples include data transformations, interactions, and techniques such as polynomial regression [62], ridge [63], and lasso regularizations [40]. Although these techniques can help increase the flexibility of regression analysis, they are applicable in specific situations and cannot be generalized for applying in autocorrelated time series and multidimensional nonlinear datasets with high variance.

Regression models have also been extensively used in short-term [46, 47, 50, 64–67] and long-term electricity forecasting [60, 68–71]. Most of the research focused on applying multiple linear regression models to forecasting electricity load [46, 47, 65, 67, 71] based on diverse weather and temporal features. Regression methods have also been explored from perspectives of short-term electricity production forecasting [48, 64], monitoring long-term consumption [60, 68, 69, 71], short- and long-term electricity price prediction problems [50, 66, 70], and operation- and maintenance-related issues [47, 72].

Despite the fact that regression models have limitations to handle nonlinear and autocorrelated data, various approaches have been investigated to extend regression models to model the random variability to a certain extent and provide reasonably accurate predictions by incorporating exponential smoothing (ES) parameters [67], Gaussian process [73], and hybrid methods [47, 74].

4.4.1.2 Exponential Smoothing and Holt-Winter's Techniques

Exponential smoothing (ES) is a well-known statistical method based on exponential window functions and traditionally has been used for forecasting in practical applications of signal processing and time series analysis [39]. The idea of exponential smoothing is similar to the moving averages technique, but the main difference is that exponential smoothing applies exponentially diminishing importance to previous observations unlike its counterpart. More specifically, it assigns higher weights to more recent observations in comparison to the older values in the sequence. As a consequence of that, the regular ES method is not capable of handling trend or seasonality in a time series.

Eventually, the Holt-Winter's exponential smoothing method, also known as Triple Exponential Smoothing, was developed as an extension to simple ES to incorporate both seasonality and trend patterns [75]. Holt-Winter's method comprises three smoothing equations, which are used to estimate seasonality, trend, and level and a forecasting equation [76] as formulated in Eqs. (4.4), (4.5), (4.6), and (4.7):

$$\text{Seasonality}: I_t = \delta\left(\frac{X_t}{S_t}\right) + (1-\gamma) I_{t-1} \tag{4.4}$$

$$\text{Trend}: T_t = \gamma\left(S_t - S_{t-1}\right) + (1-\gamma) T_{t-1} \tag{4.5}$$

$$\text{Level}: S_t = \propto\left(\frac{X_t}{I_{t-s}}\right) + (1-\propto)\left(S_{t-1} + T_{t-1}\right) \tag{4.6}$$

$$\text{Forecasting}: \hat{X}_t(\mathrm{k}) = (S_t + \mathrm{k}\,T_t)\, I_{t-s+k} \tag{4.7}$$

In Eqs. (4.4), (4.5), and (4.6), the parameters ∝, γ, and δ denote the smoothing parameters of the model and I_t, S_t, and T_t are the smoothing equations known as seasonality, trend, and level, respectively, that are estimated separately. In Eq. 4.7, $\hat{X}_t(k)$ represents the forecasting values at future step k that are made based on the observations (X_t).

Traditionally, Holt-Winter's exponential smoothing method has been frequently applied to energy forecasting [53, 54, 77–79]. Abdesselam et al. developed an approach that modifies Holt-Winter's method with Fuzzy logic. They reported improvements in the prediction accuracy when applied to the demand forecasting problem [53]. In a related study, Jónsson et al. proposed an approach for applying Holt-Winter's method to short-term estimation of electricity prices in real-time electricity markets [54].

In another related work, Hussain et al. [79] evaluated the forecasting performances of the Holt-Winter's and the ARIMA models of using electricity consumption data. They found out that the Holt-Winter's achieved better prediction accuracy compared to ARIMA model. Similarly, the studies [77, 78] explored application of Holt-Winter's to short-term electricity load demand prediction in time series data.

4.4.1.3 AR-MA and ARIMA

ARIMA stands for Autoregressive Integrated Moving Average; it is a class of statistical techniques that are primarily used for forecasting time series data. ARIMA models assume stationarity, meaning that statistical properties do not change over time. However, if the time series is not stationarity, it can be transformed to become stationarity by applying differencing and, if required, other nonlinear transformations, for example, logging or deflating [14].

ARIMA is an extended version of autoregressive-moving-average model (ARMA) that integrates autoregressive (AR) and moving-average (MA) terms in regression analysis with time series. ARIMA generalizes ARMA by adding integration term, which simply represents the differencing process that can be applied to the original series to help make it stationary [28]. ARIMA is also known as Box-Jenkins models, which is named after two prominent statisticians who developed systematic techniques for applying them to time series data for practical use [37].

ARIMA models can be generalized as formulated in Eq. 4.8. In the equation , $\hat{y}_t$ stands for the predicted value, y denotes the lags of the time series, and φ represents corresponding coefficient of the lags in AR terms, while c is the constant term that the model estimates. On the moving-average side, the ε values denote the forecast errors at the lags and θ are the estimated coefficients of those errors.

$$\hat{y}_t = c + \varphi_1 y_{t-1} + \ldots + \varphi_p y_{t-p} - \theta_1 \varepsilon_{t-1} - \ldots - \theta_q\, \varepsilon_{t-q} \tag{4.8}$$

Nonseasonal ARIMA models can be represented by three nonnegative integers and are usually represented as ARIMA(p,d,q). The symbols p, d, and q denote the model parameters, in which p stands for the quantity of autoregressive components (i.e., the number of time AR lags), d is the amount of nonseasonal differences, meaning the number of times previous values subtracted, and q means the number of the MA terms in the model.

When $d = 0$, there is no differencing, which means stationarized series is the same as the original time series. If d is equal to 1, first-order differencing is applied. Similarly, $d = 2$ stands for second-order differencing.

Identifying best possible configuration of model parameters is not an easy task. The optimal model parameters depend on the dynamic behavior of time series data. Hence, there is no universal configuration of parameters applicable to all datasets.

Although it is possible that the model parameters can be formed using a large number of combination of p, d, and q values, the model parameters generally take small nonnegative values, that is, the sum of p and q values is often less than 3. Moreover, the models frequently consist of pure AR ($q = 0$) or MA ($p = 0$) terms as some series are better fitted by AR or MA terms only.

Nonetheless, there are some intuitions that can help with selection of suitable parameters. The output of ACF and partial autocorrelation function (PACF) plots provides useful insights. PACF displays the value of autocorrelation at a time point that is not explained by lower-order autocorrelations [37, 80]. For instance, when ACF shows that values are positive at lag 1 and the values gradually decrease after initial lags and the values on PACF plot suddenly drop after a few time lags, then the series indicate existence of AR terms as shown in Fig. 4.3. In contrast, if ACF values fall sharply after a few lags and PACF values die out slowly, the time series signals presence of MA components.

Another systematic way for identifying ARIMA appropriate parameters is exploring of different parameter configurations. Evaluating the performance of models using optimization methods such as exhaustive search [81] or Random search [82] according to a loss function would reveal well-suited model parameters [83]. As a measure of relative model quality, Akaike Information Criterion (AIC) [84] is frequently utilized as the loss function in evaluation of ARIMA models.

The ARIMA algorithm has been extensively used in applications in the fields of social sciences, engineering, and finance [61, 85]. In a study, it has been observed

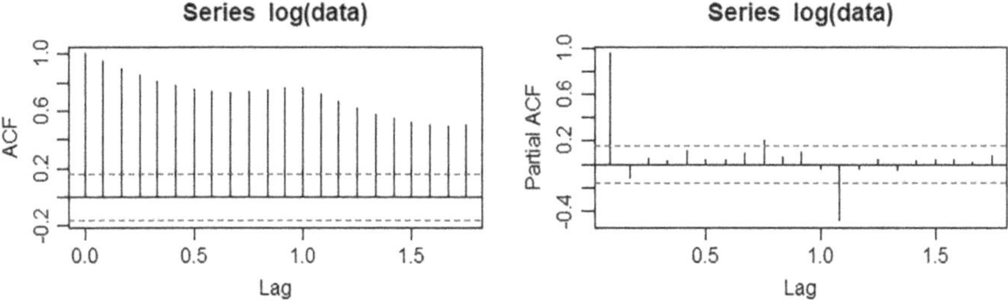

Fig. 4.3 Example ACF and PACF plots

that ARIMA was able to estimate hourly price forecast for the electricity market with an approximately 5% error rate in electricity price [51].

Another study has shown that ARIMA was able to predict electrical load estimation better than an Artificial Neural Networks (ANN) model with 20 neurons [61]. Although it resulted in better prediction performance than ANN, its prediction accuracies reduced sharply when the prediction interval increased according to the same study.

4.4.1.4 The Seasonal ARIMA and SARIMAX Models

Regular ARIMA models do not consider seasonality in the model fitting. However, sometimes, the time series may contain periodic fluctuations that require special attention in modeling. Seasonal Autoregressive Integrated Moving Average (SARIMA) is a special kind of ARIMA that is capable of modeling univariate time series data with seasonality [28].

SARIMA models extend ARIMA by including four additional seasonal parameters to existing three nonseasonal parameters. A SARIMA model can be represented by seven nonnegative integers as denoted by SARIMA(p,d,q) $(P,D,Q)m$. The symbols p, d, and q represent the nonseasonal model parameters that are the same as in nonseasonal ARIMA models. The parameters P, D, Q, and m constitute the seasonal part of the model, in which P represents the number of seasonal autoregressive terms, that is, the number of seasonal AR lags, D is the number of seasonal difference orders, Q stands for the number of seasonal MA terms in the model, and m represents the number of observations in a single seasonal period.

SARIMAX stands for Seasonal Autoregressive Integrated Moving Average Exogenous and is a type of SARIMA method that supports exogenous variables, where the character "*X*" in the method name comes from. Exogenous target means that the value of the variable is estimated externally and then integrated to the model. SARIMAX models also comprise nonseasonal and seasonal parameters as in the case of SARIMA.

Exogenous factors can have effects on the dynamic behavior of data in real world. Thus, incorporating the exogenous variables to forecasting models can be useful for realistic performance. For modeling short-term electricity production, Vagropoulos et al. compared the performances of SARIMA, modified SARIMA, SARIMAX, and ANN models [52]. They observed that SARIMA and the SARIMAX models resulted in significant performance differences. They found out that the models with exogenous factors performed better in forecasting evaluations and concluded that using exogenous factors was beneficial in a time series model.

SARIMAX models have been used in many studies for electricity forecasting in various applications [52, 55, 86–89]. The studies [86, 88] and [87] applied SARIMAX models for short-term electricity load forecasting. Similarly, Elamin and Fukushige explored applicability of SARIMAX models for estimating the demand of electricity in hourly based [55]. For the purpose of electricity price prediction, McHugh et al. also developed a SARIMAX model with 14 different external factors to predict day ahead electricity prices and achieved reasonable accurate prediction results [89].

4.4.2 *Machine Learning Methods*

As a subfield of artificial intelligence, machine learning (ML) can be roughly defined as a branch of algorithms that have the ability to learn from the data and improve automatically [90]. Machine learning constitutes algorithms that can be roughly classified into three categories such as supervised, unsupervised, and reinforcement learning [91].

Prediction methods generally belong to the supervised learning approach of machine learning that infers a mathematical model can learn patterns from a collection of training data points and is able to predict unseen records.

4.4.2.1 Individual Methods

There have been countless studies applying machine learning techniques in various contexts to forecasting problems from diverse perspectives in recent decades. From an electricity forecasting point of view, machine learning algorithms have also been increasingly extensively investigated. Constituent machine learning algorithms have been commonly used in electricity forecasting applications, including Artificial Neural Networks (ANN) in short term [56, 92, 93], long-term electricity load forecasting [94, 95] and electricity price forecasting [96, 97], support vector regression (SVR) in short term [49, 98–100], long-term energy consumption forecasting [93, 98, 101] and electricity price forecasting [102], K-Nearest Neighbors (KNN) in short term [103, 104] and long-term electricity load forecasting [56], and Decision Tree algorithms [93, 102] in electricity load forecasting.

In addition to individual machine learning algorithms, forecasting models can be developed by using ensemble methods, which employ multiple ML algorithms to achieve better prediction performance.

4.4.2.2 Ensemble Methods

In machine learning, ensemble methods build a model consisting of a combination of simpler models. Bootstrap Aggregation (Bagging), Boosting, and Stacking [105] are prominent ensemble methods.

In Bagging, a model consists of multiple base models formed by sampling from training sets. The model makes the prediction by taking the mean output of base models for the regression or majority for classification. Bagging models aim to reduce variance caused by the sampling in its estimation rather than the model bias. Random Forest algorithm is a well-known bagging technique. Researchers have regularly implemented Random Forest algorithm and reported successful prediction results in a wide spectrum of forecasting problems from real-time electricity pricing [106] to electricity load forecasting [93, 107], energy utility management [108], and electricity production [44, 109].

Alternatively, boosting is an iterative method, in which models are trained sequentially. In boosting, each model attempts to decrease the remaining error from the previous models. Contrary to bagging, the objective of boosting is to reduce bias and, subsequently, the variance in the model. XGBoost, Gradient Boosting, AdaBoost, LightGBM, and CatBoost are notable boosting ensemble algorithms. From an electricity forecasting point of view, numerous studies proposed forecasting approaches based on Boosting algorithms, which frequently provide highly successful prediction power, to broad range of practical applications including but not limited to energy planning and management in smart grids [45], short-term and long-term electricity forecasting [110–112], predicting power consumption in buildings [113] and regions [114], and electricity price forecasting [115, 116] in the literature.

Having the ability to learn patterns automatically from data, the capacity to model high-dimensional and nonlinear data, the applicability to regression problems, and providing high prediction power can be considered among the main reasons why machine learning methods have attracted a lot of attention from electricity forecasting. Despite that, machine learning algorithms are not well suited for capturing recursive and autocorrelated patterns in sequence data as is the case in time series. More advanced deep learning methods such as Recurrent Neural Networks are known to provide better learn relationships in sequence data.

4.4.3 Deep Learning Algorithms

Deep learning is a category of machine learning that can be simply explained as a learning methodology that employs multiple layers of information-processing units in a neural network to learn latent features from input data automatically and detect patterns in data [117]. The term "deep" indicates the use of large number of layers in the network. Thanks to availability of extra layers, deep learning algorithms have the ability to learn complex structures and features in data more effectively than the shallow models. However, it comes with an additional cost of training the network that requires substantially more time and computational resources to train the models. With the recent advancements in parallel processing, GPU, and cloud technologies that reduced costs for computational processing, deep learning algorithms have become increasingly available recently to widespread use in applications from image processing to language translation, time series forecasting [118].

4.4.3.1 Deep Artificial Neural Networks (Deep ANNs)

As one of the well-known deep learning methods, deep ANNs have been frequently used in short- and long-term predictions of electricity consumption [119]. When compared to ARIMA results, ANN models have been found to be more successful with using only three layers in the model. ANN's ability to learn non-linear patterns and estimate the latent associations between input features and output variable through the data has been a key factor for outperforming the ARIMA results [120].

4.4.3.2 Recurrent Neural Networks (RNNs)

Recurrent neural networks (RNNs) are a special type of artificial neural networks that allow hidden layers to be self-connected [121]. This means that the network can leverage sequential data or time series data such that the output of the previous layer is used as input to current layer along with its hidden state. This makes RNNs well suited for representing temporal dynamic effects. In RNNs, sequential information in the prior layers can propagate within the network as input at time t includes not only the current input vector but also the layer output at the previous time point $t - 1$ [122]. In other words, the predictions are influenced by combined behavior of current input and data from the recent past.

In RNNs, the standard method for training is back propagation through time (BPTT), which is an extension of back-propagation for Artificial Neural Networks [122]. The idea of BPTT is similar to regular back-propagation in feed-forward networks, except that the training is performed by unfolding the structure of the sequence and the errors also propagate within the unfolded sequence [121]. Because of that, training requires significant computational time and memory relative to regular ANNs [123]. Nevertheless, RNNs have an ability to form a very large number of connected layers to capture complex patterns in time sequences, which is critical for deep learning tasks.

Traditionally, RNNs have been used for overcoming the problem of long-term dependencies in sequence data. However, most of the RNN methods suffer from Vanishing Gradient problems, and less frequently Exploding Gradient problem. Many variants of RNN architectures have been developed such as Simple Neural Networks (SRN) [124, 125], Bidirectional Recurrent Neural Networks (BRNN) [126], Gated Recurrent Unit (GRU) [127, 128], and Long Short-Term Memory (LSTM) [129]. Particularly, GRU and LSTM algorithms have emerged as architectures to overcome the problem of Vanishing and Exploding Gradients [130].

Long Short-Term Memory (LSTM) Algorithm

The Long short-term memory (LSTM) algorithm is an advanced type of artificial recurrent neural network (RNN) that is used for capturing patterns in sequential data [129]. Unlike conventional ANN methods, LSTM algorithm can learn long-term patterns and retain hidden state while processing sequence data. As LSTM architecture has memory cells to store long-term effects, it is able to capture both short-term and long-term patterns in data effectively. LSTM fixes the Vanishing Gradient issue elegantly [131] by keeping track of long-term sequential memory using separate cell states.

Generic LSTM architecture is composed of a collection of units recurrently connected. Each LSTM neuron consists of a cell state, which memorizes long-term sequential information, and three gates that coordinate the information flowing through the cell by applying a series of activation functions and vector operations. The gates comprise an input gate, an output gate, and a forget gate.

$$f_t = \sigma\left(W_{xf} x_t + W_{hf} h_{t-1} + b_\mathrm{f}\right) \tag{4.9}$$

The forget gate determines how much of prior cell state should be kept or forgotten within the update. In Eq. 4.9, f_t represents forget gate's activation vector. It applies a sigmoid function to the input consisting of x_t, which is the input vector to the LSTM unit along with corresponding weight matrices of W_{xf} and h_{t-1} the previous cell's output plus bias vector parameters b_f to scale between [0, 1]. The output is then multiplied to the previous cell's state c_{t-1}.

$$i_t = \sigma\left(W_{\mathrm{xi}} x_t + W_{hi} h_{t-1} + b_i\right) \tag{4.10}$$

$$u_t = \tanh\left(W_{xu} x_t + W_{hu} h_{t-1} + b_u\right) \tag{4.11}$$

The input gate controls how much current cell state of the LSTM unit should be updated. Said differently, it decides whether or not to add new state information to the current cell state. Similar to the forget gate, the input gate applies a sigmoid activation function to the input vector to the LSTM unit along with corresponding weight matrices and prior cell's output to obtain activation vector i_t as formulated in Eq. 4.10. Additionally, it applies a hyperbolic tangent activation function to the input vector of the LSTM unit along with a corresponding weight to obtain cell input activation vector u_t as represented in Eq. 4.11.

$$c_t = i_t * u_t + f_t * c_{t-1} \tag{4.12}$$

As formulated in Eq. 4.12, the current state c_t is updated to the old cell state. When the activation input gate's activation value becomes very small, meaning that i_t is approximately zero, the current cell state c_t is not updated. Otherwise, it updates the old cell state.

$$o_t = \sigma\left(W_{xo} x_t + W_{ho} h_{t-1} + b_\mathrm{o}\right) \tag{4.13}$$

$$h_t = \tanh\left(c_t\right).^{*} o_t \tag{4.14}$$

The output gate decides how much hidden state of the current LSTM unit should be passed to the next LSTM unit. In other words, it applies a sigmoid activation to the input of the LSTM unit and previous cell's output to obtain output vector o_t as formulated in Eq. 4.13. Then, the hyperbolic tangent activation of cell state c_t is multiplied by output vector o_t to calculate the output h_t of LSTM unit as denoted in Eq. 4.14.

As LSTM networks have the ability to learn long-term dependencies in sequence data and time series data resemble lags of events as order-dependent sequence of time, they are an ideal candidate for prediction in deep learning tasks using time series data.

It has been successfully used in many applications in the field. There are studies achieving 60–65% accuracies in predicting stock price by trying various optimization techniques [132]. Bouktif et al. have shown that a hybrid LSTM model gave successful results in the estimation of energy load [57].

More comprehensive models such as Sequence to Sequence approach used in combination with RNN S2S or LSTM S2S appeared to outperform traditional RNN (Recurrent Neural Networks) or LSTM models [133]. In another study, Marino et al. have also observed that using the classical LSTM method combined with LSTM's S2S (Sequence to Sequence) architecture produced successful results in their evaluations [134].

In another study, it aimed to produce realistic estimates using deep learning methods [135]. Daily electricity consumption predictions were made using a time series dataset containing hourly electricity consumption for a year. Appropriate preprocessing tasks have been completed in the first phase of analysis. In the modeling phase, LSTM and GRU algorithms, two of the most popular methods of deep learning, and ARIMA method, one of the most frequently used time-series analysis techniques, were evaluated in a comparative manner.

During the evaluations, the hyperparameter tuning was performed for each method to find their most suitable parameters that produce the best prediction results. Based on the comparison results, it was noticed that the best prediction accuracy was obtained using LSTM with an error rate of 13%. Although the method GRU also performed well and had a close error rate, it still lagged behind LSTM. On the other hand, ARIMA was insufficient for the forecasting when compared to the other two deep learning methods [135].

Gated Recurrent Unit (GRU) Algorithm

The Gated Recurrent Unit (GRU) algorithm is a more recent and sophisticated version of regular Recurrent Neural Networks (RNNs). Proposed by Cho et al. in 2014, GRU offers a gating mechanism to avoid Vanishing Gradient problem [127]. In standard architecture, GRU comprises a set of recurrently connected units, in which each unit consists of two gates, namely, reset gate and update gate.

GRU and LSTM algorithms have similar learning mechanisms but the former has a less complex architecture. Unlike LSTM, GRU lacks an output gate and does not retain a separate cell state. Thus, it combines the input vector from its current unit and the hidden state from prior unit and passes the entire hidden state as an output to the next unit in GRU architecture.

$$z_t = \sigma\left(W_z x_t + U_z h_{t-1} + b_z\right) \tag{4.15}$$

The update gate controls how much of prior information should be passed or thrown away within the update. In Eq. 4.15, z_t represents update gate's activation vector. It applies a sigmoid function to input consisting of x_t, which is the input vector to the GRU cell along with corresponding weight matrices of W_z and h_{t-1} previous cell's output as well as its weights U_z plus bias vector parameters b_z to scale between [0, 1]. The output is then multiplied to previous cell's state c_{t-1}. This operation at the update gate can help prevent the possibility of vanishing gradient problem, as it can pass along all the state from the previous units.

$$r_t = \sigma\left(W_r x_t + U_r h_{t-1} + b_r\right) \tag{4.16}$$

The reset gate controls the long-term behavior of the network and determines how much of prior information should be forgotten within the update. In Eq. 4.16, r_t represents the reset gate's activation vector. The gate applies a sigmoid function to the input vector to the GRU unit along with corresponding weight matrices and prior cell's output to obtain activation vector r_t.

$$h_t' = \tanh\left(W_h x_t + U_h\left(r_t \odot h_{t-1}\right) + b_h\right) \tag{4.17}$$

The reset gate manages the short-term memory of the network by influencing the amount of state to pass onto the candidate hidden state h'_t. The candidate hidden state is calculated by applying a hyperbolic tangent activation function to the input vector of the current GRU unit along with corresponding weights plus multiplying the weight parameters U_h to the output of Hadamard product [136, 137], which is denoted by operator $\odot$ between reset gate's activation vector and previous hidden state as represented in Eq. 4.17.

$$h_t = z_t \odot h_{t-1} + \left(1 - z_t\right) \odot h'_t \tag{4.18}$$

When the candidate hidden state is computed, it can be used for evaluating the hidden state h_t of the current unit to be passed to the next unit. Different from LSTM architecture, GRU does not possess a separate gate for transferring output state; instead, it employs the update gate to serve for this purpose. As expressed in Eq. 4.18, the output of update gate balances how much candidate hidden state of the current GRU unit and the hidden state from the prior unit should influence the memory to be transmitted to the next unit. This is a critical step for resolving the potential occurrence of a vanishing gradient problem.

It is because the recurrent unit may or may not pass along the entire hidden memory from the previous units. The models can learn how to filter prior memory and current hidden state, and then pass on the relevant information to the next units in the network. When z_t, the output of update gate, becomes 0, the hidden state h_t of the current unit will be entirely based on the candidate state at the current unit and will not transmit memory from previous units. Contrarily, if z_t approaches the value of 1, the output of current hidden state will be affected by the memory from prior units only.

The main advantage of GRU is that the model training time using GRU is generally shorter and it is less memory intensive than LSTM as the GRU architecture is simpler and contains less parameters to train than the latter [127]. Although it is less complex than LSTM, the performance of GRU is often comparable to the LSTM's.

In some studies, GRU models implemented with various optimizations appeared to have similar or slightly better performance in comparison to other machine learning algorithms in cases where relatively close results are obtained [138]. In another study, Sajjad et al. applied a GRU model for prediction of heating and cooling loads [139].

4.4.3.3 Convolutional Neural Networks (CNNs)

Convolutional Neural Networks (CNNs) are in essence a type of Deep Artificial Neural Networks. The CNN models additionally have a set of convolutional layers that are particularly useful for extraction of features from raw input on top of a typical Artificial Neural Network [118]. The features that are extracted using convolutional layers are then flattened and subsequently used as input to regular fully connected neural networks for model training.

Although CNN models have been particularly applied to image- and video-processing problems, there have been innovative studies exploring the applicability of CNNs to the electrical energy prediction in time series data [58, 59, 140]. For instance, some hybrid deep learning approaches have been proposed that extract high-level spatial features using CNN layers and then feed these features into LSTM network to learn patterns of long-term sequences in the time series data [59].

4.5 Hyperparameter Optimizations and Evaluation Metrics

4.5.1 Hyperparameter Tuning

The optimization phase in machine learning and deep learning methods is a challenging task. From both dynamic and objective perspectives, different configurations need to be explored to accomplish the forecasting modeling task effectively. Said differently, it is often necessary to evaluate various combinations of configuration parameters (a.k.a. hyperparameters) to achieve peak performance of machine learning models. Optimal selection of these hyperparameters can significantly increase the overall prediction performance of the model.

For systematic exploration of optimal hyperparameters in models, several hyperparameter optimization methods have been developed that estimate the performances of different models based on a predefined loss function [83]. These optimization methods typically employ a cross-validation technique to reduce variance introduced from sampling and generalize the performance evaluations [81, 82].

Grid search [81], Bayesian optimization [141], evolutionary optimization [81], and random search [82] are among the most commonly utilized tuning methods in testing the performance of machine models. On the other hand, the choice of optimization method as one of the algorithm's hyperparameters can have a huge impact on the overall model performance. The Grid search method exhaustively evaluates combinations of model hyperparameters and thus can lead to an excessive use of computation time and resources. For this reason, Random search, which selectively evaluates hyperparameter combinations and provides comparative results, is often preferred over the Grid search technique.

When optimizing deep learning algorithms, for example, LSTM and GRU, the set of hyperparameters must be carefully selected to yield an optimal model. The batch size, the number of neurons and epochs, optimization method, loss function, and learning rate can be considered among the main parameters. In a study by Chang et al., optimization methods of the LSTM neural networks were compared from electricity price forecasting perspective [142]. They found out that the Adaptive Moment Estimation (ADAM) optimization method appeared to have more accurate outcomes applications in the forecasting the electricity price when compared to Stochastic Gradient Descent (SGD) and Root Mean Square Spread (RMSProp) optimization methods.

4.5.2 Evaluation Metrics

In regression and time series analysis, model performances must be evaluated using model quality measures that are able to estimate prediction errors in terms of the total sum of differences between the predictions and actual observations in a quantitative manner. While some measures calculate the error of predictions in absolute terms, others compute the loss in form of proportions. Model evaluation metrics including Root Mean Squared Error (RMSE), Mean Absolute Error (MAE), Mean Absolute Percentage Error (MAPE), and R-squared (R^2) are among the most commonly used in electricity forecasting applications.

Root Mean Squared Error (RMSE) is a widely utilized error measure that gauges the standard deviation of the prediction errors from actual observations [143, 144]. The smaller the model's RMSE value is the better, as this means the model predictions are fitting the actual values well. As shown in Eq. 4.19, the RMSE metric is defined as follows. In the formula, $\hat{y}_i$ refers to the predicted values, whereas y_i denotes the actual observations.

$$\text{RMSE} = \sqrt{\frac{\sum_{i=1}^{n}\left(y_i - \hat{y}_i\right)^2}{n}} \tag{4.19}$$

Mean Absolute Error (MAE) is another commonly used metric that estimates the average difference between the model predictions. Equation 4.20 demonstrates the formula used for calculating the MAE. MAE measures the deviances of predictions from the actual failures in the test set in terms of absolute values [145].

$$MAE = \frac{\sum_{i=1}^{n}\left|y_i - \hat{y}_i\right|}{n} \tag{4.20}$$

Similar to MAE, Mean Absolute Percentage Error (MAPE) measures the total deviation of model predictions from the actual values, but the error term is represented in terms of a ratio instead of an absolute value [146]. The smaller values of both MAE and MAPE indicate the better model fitting. The formula of the metric is expressed in Eq. 4.21 as follows:

$$\text{MAPE} = \frac{100}{n}\sum_{i=1}^{n}\left|\frac{y_i - \hat{y}_i}{y_i}\right| \tag{4.21}$$

R-squared (R^2) is another well-known metric that measures overall quality of model fitting. It represents the proportion of overall uncertainty that the model removed. Said differently, R^2 estimates the ratio of the total variability in the target variable that was reduced after fitting the model. The highest value R^2 can take is 1, which represents perfect fit to the observations. The higher R^2 value means the model is fitting better [41]. Equation 4.22 illustrates the formula of the calculation.

$$R^2 = 1 - \frac{\sum_{i=1}^{n}\left(y_i - \hat{y}_i\right)^2}{\sum_{i=1}^{n}\left(y_i - \bar{y}_i\right)^2} \tag{4.22}$$

4.6 Applying AI Algorithms to Forecast Power Consumption: A Use Case

In this chapter, a use case was implemented based on a real-world dataset to evaluate the performances of the aforementioned artificial intelligence techniques in electricity forecasting. The following subsections describe the details of the use-case application.

4.6.1 Data Collection and Preprocessing

The dataset was collected from an electricity distribution company in Turkey. It comprised electricity consumption of a region that recorded per hour for the duration of a year. The variables consisted of the total power consumption, temperature, and humidity in the region along with the timestamp of recording.

As none of the features in the dataset contained missing values, there was no need for applying data filling techniques. Apart from the timestamp of the observation, all features were numerical. When preprocessing the dataset, it was observed that there was practically no trend in the features over the time period. However, the power consumption presented weekly and daily cycles as it can be seen clearly in Fig. 4.4.

Another observation from the data explorations that drew our attention was a long duration of low level of energy consumption. This period can be seen in Fig. 4.4 around time points between 650 and 850. This phenomenon appears to be related to a large-scale power outage in that particular region rather than an outlier or noise in data. Thus, these records were not removed from the data analysis.

The ranges of the features in the dataset varied considerably. Therefore, data normalization was applied to bring all features to the scale between 0 and 1 during preprocessing phase. This is a necessary process in machine learning to help reduce the sensitivity of models to different scales and enhance the general quality of the models.

Moreover, the dataset was split into training and test set; 80% of the dataset was used for training and the remaining held out for testing. Keeping a separate test dataset for validation is necessary for discovering potential overfitting and underfitting problems in modeling.

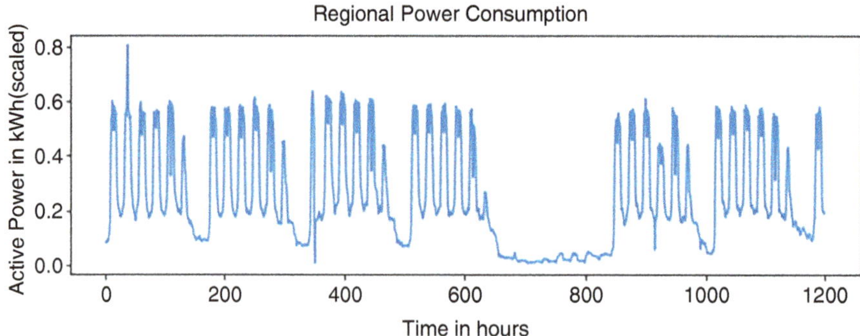

Fig. 4.4 An excerpt from the power consumption dataset

4.6.2 Developing AI Models

In the use case, the prediction of power consumption can be considered as a regression problem. However, the observation points in the dataset were time dependent and represented in form of time series. Thus, the algorithms in the evaluations were chosen among the methods that have the ability to model dynamic patterns in sequential data.

Six different algorithms were selected for performing comparative evaluations in the study. The methods utilized in the analysis consisted of two deep learning algorithm, one statistical time series analysis technique, and three different machine learning methods. The RNN-based deep learning algorithms were, namely, LSTM and GRU. Additionally, the machine learning algorithms used in the assessment comprised Random Forests, Artificial Neural Network (ANN), and Multiple Linear Regression algorithms. As a statistical time series analysis technique, ARIMA was also selected for the performance evaluations.

In the analysis, numerous hyperparameters were explored for identifying best performing models for each algorithm. Based on various hyperparameters options for each model, the following hyperparameters were found optimal by using randomized search method.

For the LSTM model, the most successful results were obtained when choosing 100 epochs, 72 batch size, and 50 × 50 neuron size with ADAM optimization method. Differently on GRU, the best hyperparameters were 100 epochs, 16 batch size, 100 × 200 neurons, and ADAM optimization among the tested configurations. For ARIMA, the best performing model was obtained when setting (p,d,q) parameters to (2, 0, 1), respectively. Similarly, the number of estimators of 51 and minimum sample split of 2 were found to be performing well when training the Random Forests model. In the case of ANN model, the hyperparameters of three hidden layers with 4 × 64 × 64 nodes, ADAM optimizer, and 100 epochs with hyperbolic tangent activation function appeared to work well among various tuning options.

Deep learning models are prone to have overfitting issues. Thus, it is necessary to ensure that the models are not overfitting the data in training. We checked the presence of potential model fitting problems in the evaluations before moving forward to the next steps. As shown in Fig. 4.5, the training and testing losses were consistent and converged to approximately the same values right after 20 epochs in both cases of GRU and LSTM models. There was no noticeable gap between the training and testing results. This means that overfitting was not observed for these models.

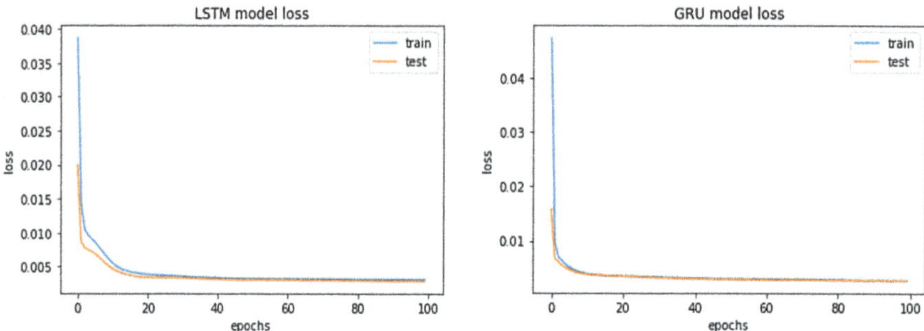

Fig. 4.5 Fitting LSTM and GRU models

Table 4.1 Performance evaluation results of the models

Algorithms	RMSE	MAE	MAPE
LSTM	0.053	0.033	13.34
GRU	0.053	0.034	13.44
ARIMA	0.058	0.037	14.65
ANN	0.177	0.042	16.79
Random Forests	0.126	0.097	34.31
Linear Regression	0.182	0.155	55.21

4.6.3 *Evaluating the Model Results*

In the evaluations, the prediction performances of the algorithms were compared using the metrics of Root Mean Squared Error (RMSE), Mean Absolute Error (MAE), and Mean Absolute Percentage Error (MAPE). Table 4.1 demonstrates the evaluation results on the testing dataset.

The best performing models were obtained when using deep learning models. Although LSTM achieved the top performance with 0.053 RMSE, 0.033 MAE, and 13.34 MAPE values, its differences from GRU model were negligibly small and RMSE scores were identical for both models. As expected, the performance of ARIMA model lagged behind of those aforementioned RNN models.

Contrarily, the machine learning models using ANN, random forests, and linear regression algorithms performed poorly when compared to the three sequential data modeling methods above and found to be ineffective for modeling time series data in the use case. The worst performance was obtained clearly when using multiple linear regression model with 0.182 RMSE, 0.155 MAE, and 55.21 MAPE scores.

In the evaluations, the prediction performances of these six models on the test dataset were also visually explored. Figure 4.6 presents an excerpt from the observations and predictions of evaluated models for demonstration purposes. As shown in the figure, it is clear that the LSTM and GRU models were the most successful in prediction of the power consumption. They were closely followed by the ARIMA model. On the other hand, ANN, Random Forests, and Multiple Linear Regression

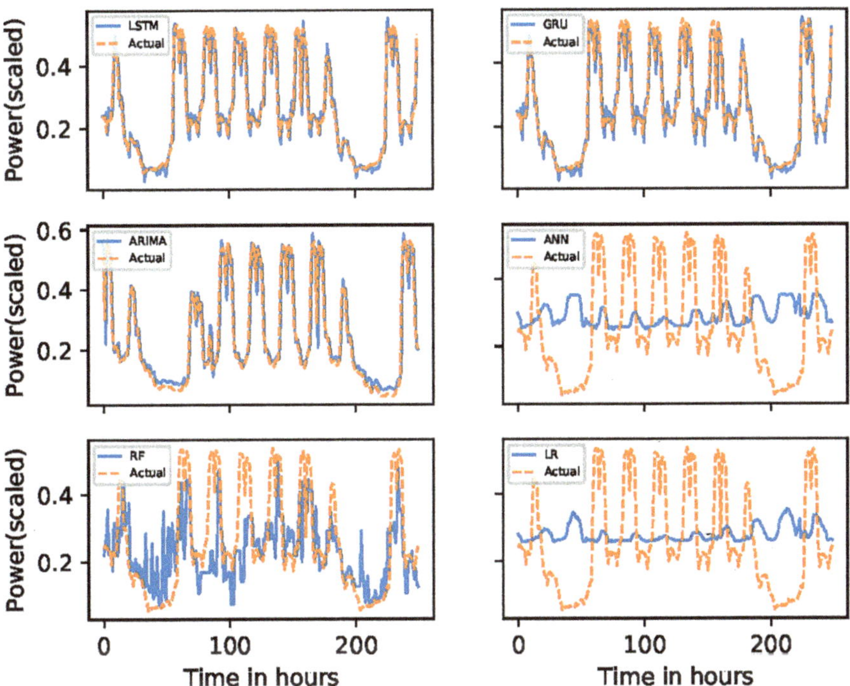

Fig. 4.6 Evaluation results of the algorithms with respect to actual vs. predicted values

algorithms were unable to capture the dynamic patterns over the time period. Although it may not be obvious from the results of error measures, it is more visible on the figure that the Random Forests model achieved considerably better in fitting the data than the ANN model.

4.6.4 Discussions

Electricity forecasting plays a crucial role in a wide spectrum of applications in energy sector. The stakeholders in the electricity market heavily rely on the information provided by the forecasting models when making short-term operational decisions and planning long-term strategies. To produce useful insights for decision making and become effective, the forecasting models need to be able to capture dynamic patterns in the data and minimize the prediction errors.

In the chapter, a use case was developed as a realistic simulation to demonstrate the applicability of artificial intelligence methods to the short-term electricity demand forecasting. The selection and optimization of algorithms is the key to the success of practical forecasting applications. In the use-case analysis, the forecasting performances of several methods were assessed using a real-world dataset. Based on the results of the analysis, the deep RNN algorithms of LSTM and GRU produced the best forecasting results. Despite the fact that the LSTM model achieved marginally better results, the GRU model can be considered as the best model in the evaluations when taking into account the additional computational overhead generated for training the LSTM model.

Another important observation was that as a commonly used statistical time series analysis method, ARIMA model was not quite successful in comparison to the deep RNN methods. This may not surprising, since ARIMA performs Univariate data analysis and is not suitable for adapting dynamic variations in multivariate data as opposed to the RNNs. Having said that, the ARIMA model still produced significantly better results when compared to the Random Forests, ANN, and Multiple Linear Regression models. Yet this is a reasonable outcome for the analysis because the latter three models predict the target variable based on the input variables and treat it as a regression problem. Those regression models are not suitable for fitting the temporal sequences presented in time series data as they do not consider the dependencies between the input record and previous observations.

Moreover, the independent variables of humidity and temperature solely do not seem to be sufficient for the prediction of power consumption in regression analysis. It is conceivable that these regression models could potentially produce better forecasting results, if additional independent factors in relation to power consumption were available in the dataset.

One limitation of the dataset in the use case was the duration of the time interval. The dataset consisted observations from 1 year only. This restricted the analysis of the changes in power consumption behaviors over multiple years. Consequently, there was no need to apply a large number of layers for training the deep learning models as the number of observations in the dataset was low.

4.7 Conclusions

Effectively regulating electricity production and demand is of great importance and has been an increasing need over time. Because of the fact that electricity is a type of energy that is nonstorable and the amount of consumption is continuously increasing, accurate forecasts are essential in management of smart grids, calculating prices, estimating capacity and demand, assessing feasibility, and planning new investments. In this book chapter, short-term and long-term electricity forecasting methods ranging from statistical approaches to machine learning and deep learning algorithms were reviewed in depth within the context of artificial intelligence. Furthermore, a variety of notable applications and methods were discussed. The chapter also presented the challenges and conditions in electricity forecasting along with the fundamental properties of time series data. In conclusion, the techniques of hyperparameter optimization and evaluation metrics were described in detail.

References

1. Z.H. Gontar, *Smart Grid Analytics for Sustainability and Urbanization* (IGI Global, 2018)
2. M. Rahmani-Andebili, Cooperative distributed energy scheduling in microgrids, in *Electric Distribution Network Management and Control*, (Springer, 2018), pp. 235–254
3. M. Rahmani-Andebili, Stochastic, adaptive, and dynamic control of energy storage systems integrated with renewable energy sources for power loss minimization. Renew. Energy **113**, 1462–1471 (2017)
4. M. Rahmani-Andebili, Dynamic and adaptive reconfiguration of electrical distribution system including renewables applying stochastic model predictive control. IET Gener. Transm. Distrib. **11**(16), 3912–3921 (2017)
5. F. Caputo, B. Buhnova, L. Walletzkỳ, Investigating the role of smartness for sustainability: Insights from the smart grid domain. Sustain. Sci. **13**(5), 1299–1309 (2018)
6. M. Faheem et al., Smart grid communication and information technologies in the perspective of Industry 4.0: Opportunities and challenges. Comput. Sci. Rev. **30**, 1–30 (2018)
7. T.-H. Dang-Ha, R. Olsson, H. Wang, The role of big data on smart grid transition, in *2015 IEEE International Conference on Smart City/SocialCom/SustainCom (SmartCity)*, (2015), pp. 33–39
8. C. Müller et al., Modeling framework for planning and operation of multi-modal energy systems in the case of Germany. Appl. Energy **250**, 1132–1146 (2019)
9. P. Mancarella, MES (multi-energy systems): An overview of concepts and evaluation models. Energy **65**, 1–17 (2014)
10. A. Asrari, S. Lotfifard, M. Ansari, Reconfiguration of smart distribution systems with time varying loads using parallel computing. IEEE Trans. Smart Grid **7**(6), 2713–2723 (2016)
11. C.T. Larsen, G. Doorman, B. Mo, Joint modelling of wind power and hydro inflow for power system scheduling. Energy Procedia **87**, 189–196 (2016)
12. P. Girardi, A. Temporelli, Smartainability: A methodology for assessing the sustainability of the smart city. Energy Procedia **111**, 810–816 (2017)
13. G. Dileep, A survey on smart grid technologies and applications. Renew. Energy **146**, 2589–2625 (2020)
14. K. Gajowniczek, T. Ząbkowski, Short term electricity forecasting using individual smart meter data. Procedia Comput. Sci. **35**, 589–597 (2014)
15. S. Khatoon, A.K. Singh, others, Effects of various factors on electric load forecasting: An overview, in *2014 6th IEEE Power India International Conference (PIICON)* (2014), pp. 1–5
16. E.M. Eskandarnia, S.A. Kareem, H.M. Al-Ammal, A review of smart meter load forecasting techniques: Scale and horizon, (2018)
17. E. Mocanu, P.H. Nguyen, M. Gibescu, W.L. Kling, Deep learning for estimating building energy consumption. Sustain. Energy Grids Netw. **6**, 91–99 (2016)
18. X. Wang, F. Fang, X. Zhang, Y. Liu, L. Wei, Y. Shi, LSTM-based short-term load forecasting for building electricity consumption, in *2019 IEEE 28th International Symposium on Industrial Electronics (ISIE)*, (2019), pp. 1418–1423
19. J. Massana, C. Pous, L. Burgas, J. Melendez, J. Colomer, Short-term load forecasting in a non-residential building contrasting models and attributes. Energ. Buildings **92**, 322–330 (2015)
20. Z. Wang, R.S. Srinivasan, A review of artificial intelligence based building energy use prediction: Contrasting the capabilities of single and ensemble prediction models. Renew. Sust. Energ. Rev. **75**, 796–808 (2017)
21. S. Sulaiman, P.A. Jeyanthy, D. Devaraj, S.S. Mohammed, K. Shihabudheen, Smart meter data analytics for load prediction using extreme learning machines and artificial neural networks, in *2019 IEEE International Conference on Clean Energy and Energy Efficient Electronics Circuit for Sustainable Development (INCCES)*, (2019), pp. 1–4
22. K. Li, C. Hu, G. Liu, W. Xue, Building's electricity consumption prediction using optimized artificial neural networks and principal component analysis. Energ. Buildings **108**, 106–113 (2015)

23. F. Cavallaro, Electric load analysis using an artificial neural network. Int. J. Energy Res. **29**(5), 377–392 (2005)
24. S. Rahman, O. Hazim, A generalized knowledge-based short-term load-forecasting technique. IEEE Trans. Power Syst. **8**(2), 508–514 (1993)
25. T. Hong, S. Fan, Probabilistic electric load forecasting: A tutorial review. Int. J. Forecast. **32**(3), 914–938 (2016)
26. K. Zor, O. Timur, A. Teke, A state-of-the-art review of artificial intelligence techniques for short-term electric load forecasting, in *2017 6th International Youth Conference on Energy (IYCE)*, (2017), pp. 1–7
27. J. Xie, T. Hong, J. Stroud, Long-term retail energy forecasting with consideration of residential customer attrition. IEEE Trans. Smart Grid **6**(5), 2245–2252 (2015)
28. R.J. Hyndman, G. Athanasopoulos, *Forecasting: Principles and Practice* (OTexts, 2018)
29. D. Gerbing, *Time Series Components* (Portland State University, 2016), p. 9
30. J.D. Hamilton, *Time Series Analysis* (Princeton University Press, 2020)
31. P.J. Brockwell, R.A. Davis, *Introduction to Time Series and Forecasting* (Springer, 2016)
32. D. Kwiatkowski, P.C. Phillips, P. Schmidt, Y. Shin, Testing the null hypothesis of stationarity against the alternative of a unit root: How sure are we that economic time series have a unit root? J. Econ. **54**(1–3), 159–178 (1992)
33. D.A. Dickey, W.A. Fuller, Distribution of the estimators for autoregressive time series with a unit root. J. Am. Stat. Assoc. **74**(366a), 427–431 (1979)
34. P.C. Phillips, P. Perron, Testing for a unit root in time series regression. Biometrika **75**(2), 335–346 (1988)
35. G. Elliott, T.J. Rothenberg, J.H. Stock, *Efficient Tests for an Autoregressive Unit Root* (National Bureau of Economic Research, Cambridge, Mass., USA, 1992)
36. E. Holmes, E. Ward, *Applied Time Series Analysis for Fisheries and Environmental Sciences* (NOAA Fish, Seattle, 2019)
37. G.E. Box, G.M. Jenkins, G.C. Reinsel, G.M. Ljung, *Time Series Analysis: Forecasting and Control* (John Wiley & Sons, 2015)
38. D.C. Montgomery, C.L. Jennings, M. Kulahci, *Introduction to Time Series Analysis and Forecasting* (John Wiley & Sons, 2015)
39. "NIST/SEMATECH e-Handbook of Statistical Methods." National Institute of Standards and Technology, 2012. Accessed: 8 Jan 2021. [Online]. Available: http://www.itl.nist.gov/div898/handbook/
40. R. Tibshirani, Regression shrinkage and selection via the lasso. J. R. Stat. Soc. Ser. B Methodol. **58**(1), 267–288 (1996)
41. N.R. Draper, H. Smith, *Applied Regression Analysis*, vol 326 (John Wiley & Sons, 1998)
42. H. Deng, G. Runger, Feature selection via regularized trees, in *The 2012 International Joint Conference on Neural Networks (IJCNN)*, (2012), pp. 1–8. https://doi.org/10.1109/IJCNN.2012.6252640
43. H. Hotelling, Analysis of a complex of statistical variables into principal components. J. Educ. Psychol. **24**(6), 417–441 (1933). https://doi.org/10.1037/h0071325
44. M. Zolfaghari, M.R. Golabi, Modeling and predicting the electricity production in hydropower using conjunction of wavelet transform, long short-term memory and random forest models. Renew. Energy **170**, 1367–1381 (2021). https://doi.org/10.1016/j.renene.2021.02.017
45. T. Ahmad, H. Zhang, B. Yan, A review on renewable energy and electricity requirement forecasting models for smart grid and buildings. Sustain. Cities Soc. **55**, 102052 (2020). https://doi.org/10.1016/j.scs.2020.102052
46. T. Hong, M. Gui, M.E. Baran, H.L. Willis, Modeling and forecasting hourly electric load by multiple linear regression with interactions, in *IEEE PES General Meeting*, (2010), pp. 1–8
47. B. Yildiz, J.I. Bilbao, A.B. Sproul, A review and analysis of regression and machine learning models on commercial building electricity load forecasting. Renew. Sust. Energ. Rev. **73**, 1104–1122 (2017)

48. M. Zamo, O. Mestre, P. Arbogast, O. Pannekoucke, A benchmark of statistical regression methods for short-term forecasting of photovoltaic electricity production. Part II: Probabilistic forecast of daily production. Sol. Energy **105**, 804–816 (2014)
49. Y. Yang, J. Che, C. Deng, L. Li, Sequential grid approach based support vector regression for short-term electric load forecasting. Appl. Energy **238**, 1010–1021 (2019)
50. T. Ulgen, G. Poyrazoglu, Predictor analysis for electricity price forecasting by multiple linear regression, in *2020 International Symposium on Power Electronics, Electrical Drives, Automation and Motion (SPEEDAM)*, (2020), pp. 618–622
51. J. Contreras, R. Espinola, F. Nogales, A.J. Conejo, ARIMA models to predict next-day electricity prices. IEEE Power Eng. Rev. **22**(9), 57–57 (2002)
52. S.I. Vagropoulos, G. Chouliaras, E.G. Kardakos, C.K. Simoglou, A.G. Bakirtzis, Comparison of SARIMAX, SARIMA, modified SARIMA and ANN-based models for short-term PV generation forecasting, in *2016 IEEE International Energy Conference (ENERGYCON)*, (2016), pp. 1–6
53. M. Abdesselam, A. Karim, H. Emrul Kays, R. Sarker, Forecasting demand: Development of a fuzzy growth adjusted Holt-Winters approach. Adv. Mater. Res. **903**, 402–407 (2014)
54. T. Jónsson, P. Pinson, H.A. Nielsen, H. Madsen, Exponential smoothing approaches for prediction in real-time electricity markets. Energies **7**(6), 3710–3732 (2014)
55. N. Elamin, M. Fukushige, Modeling and forecasting hourly electricity demand by SARIMAX with interactions. Energy **165**, 257–268 (2018)
56. H. Sangrody, N. Zhou, S. Tutun, B. Khorramdel, M. Motalleb, M. Sarailoo, Long term forecasting using machine learning methods, in *2018 IEEE Power and Energy Conference at Illinois (PECI)*, (2018), pp. 1–5
57. S. Bouktif, A. Fiaz, A. Ouni, M.A. Serhani, Optimal deep learning lstm model for electric load forecasting using feature selection and genetic algorithm: Comparison with machine learning approaches. Energies **11**(7), 1636 (2018)
58. W. He, Load forecasting via deep neural networks. Procedia Comput. Sci. **122**, 308–314 (2017)
59. M. Ishaq, S. Kwon, others, Short-term energy forecasting framework using an ensemble deep learning approach. IEEE Acccss (2021)
60. D. Angelopoulos, J. Psarras, Y. Siskos, Long-term electricity demand forecasting via ordinal regression analysis: The case of Greece, in *2017 IEEE Manchester PowerTech*, (2017), pp. 1–6
61. G. Gao, K. Lo, F. Fan, Comparison of ARIMA and ANN models used in electricity price forecasting for power market. Energy Power Eng. **9**(4B), 120–126 (2017)
62. E. Ostertagová, Modelling using polynomial regression. Procedia Eng. **48**, 500–506 (2012)
63. A.E. Hoerl, R.W. Kennard, Ridge regression: Biased estimation for nonorthogonal problems. Technometrics **12**(1), 55–67 (1970)
64. M. Zamo, O. Mestre, P. Arbogast, O. Pannekoucke, A benchmark of statistical regression methods for short-term forecasting of photovoltaic electricity production, part I: Deterministic forecast of hourly production. Sol. Energy **105**, 792–803 (2014)
65. J. Hinman, E. Hickey, Modeling and forecasting short-term electricity load using regression analysis. Inst. Regul. Policy Stud., Normal, IL, USA, Tech. Report. 1–51 (2009)
66. B. Uniejewski, R. Weron, Regularized quantile regression averaging for probabilistic electricity price forecasting. Energy Econ. **95**, 105121 (2021)
67. A.D. Papalexopoulos, T.C. Hesterberg, A regression-based approach to short-term system load forecasting. IEEE Trans. Power Syst. **5**(4), 1535–1547 (1990)
68. V. Bianco, O. Manca, S. Nardini, Electricity consumption forecasting in Italy using linear regression models. Energy **34**(9), 1413–1421 (2009)
69. V. Bianco, O. Manca, S. Nardini, Linear regression models to forecast electricity consumption in Italy. Energy Sources Part B Econ. Plan. Policy **8**(1), 86–93 (2013)
70. F. Ziel, R. Steinert, Probabilistic mid-and long-term electricity price forecasting. Renew. Sust. Energ. Rev. **94**, 251–266 (2018)
71. G. Nalcaci, A. Özmen, G.W. Weber, Long-term load forecasting: Models based on MARS, ANN and LR methods. Cent. Eur. J. Oper. Res. **27**(4), 1033–1049 (2019)

72. S. Katipamula, T.A. Reddy, D.E. Claridge, Multivariate regression modeling, (1998)
73. A. Zeng, H. Ho, Y. Yu, Prediction of building electricity usage using Gaussian process regression. J. Build. Eng. **28**, 101054 (2020)
74. A. Bello, D.W. Bunn, J. Reneses, A. Muñoz, Medium-term probabilistic forecasting of electricity prices: A hybrid approach. IEEE Trans. Power Syst. **32**(1), 334–343 (2016)
75. P.R. Winters, Forecasting sales by exponentially weighted moving averages. Manag. Sci. **6**(3), 324–342 (1960)
76. O. Trull, J.C. García-Díaz, A. Troncoso, Initialization methods for multiple seasonal Holt–Winters forecasting models. Mathematics **8**(2), 268 (2020)
77. V. Lepojevic, M. Andelkovic-Pesic, Forecasting electricity consumption by using holt-winters and seasonal regression models. Econ. Organ. **8**(4), 421–431 (2011)
78. W. Sulandari, H. Utami, others, Forecasting electricity load demand using hybrid exponential smoothing-artificial neural network model. Int. J. Adv. Intell. Inform. **2**(3), 131–139 (2016)
79. A. Hussain, M. Rahman, J.A. Memon, Forecasting electricity consumption in Pakistan: The way forward. Energy Policy **90**, 73–80 (2016)
80. P.J. Brockwell, R.A. Davis, *Time Series: Theory and Methods* (Springer Science & Business Media, 2009)
81. J. Bergstra, R. Bardenet, Y. Bengio, B. Kégl, Algorithms for hyper-parameter optimization. Adv. Neural Inf. Process. Syst. **24,** 1–9 (2011)
82. J. Bergstra, Y. Bengio, Random search for hyper-parameter optimization. J. Mach. Learn. Res. **13**(2), 281–305 (2012)
83. M. Feurer, F. Hutter, Hyperparameter optimization, in *Automated Machine Learning*, (Springer, Cham, 2019), pp. 3–33
84. H. Akaike, Information theory and an extension of the maximum likelihood principle, in *Selected Papers of Hirotugu Akaike*, (Springer, 1998), pp. 199–213
85. Ü.Ç. Büyükşahin, Ş. Ertekin, Improving forecasting accuracy of time series data using a new ARIMA-ANN hybrid method and empirical mode decomposition. Neurocomputing **361**, 151–163 (2019)
86. S. Bercu, F. Proïa, A SARIMAX coupled modelling applied to individual load curves intraday forecasting. J. Appl. Stat. **40**(6), 1333–1348 (2013)
87. A. Tarsitano, I.L. Amerise, Short-term load forecasting using a two-stage sarimax model. Energy **133**, 108–114 (2017)
88. F. Sheng, L. Jia, Short-term load forecasting based on SARIMAX-LSTM, in *2020 5th International Conference on Power and Renewable Energy (ICPRE)*, (2020), pp. 90–94
89. C. McHugh, S. Coleman, D. Kerr, D. McGlynn, Forecasting day-ahead electricity prices with a SARIMAX model, in *2019 IEEE Symposium Series on Computational Intelligence (SSCI)*, (2019), pp. 1523–1529
90. E. Alpaydin, *Introduction to Machine Learning* (MIT Press, 2020)
91. C.M. Bishop, *Pattern Recognition and Machine Learning*, (Springer, New York, NY, USA, 2006)
92. P.-H. Kuo, C.-J. Huang, A high precision artificial neural networks model for short-term energy load forecasting. Energies **11**(1), 213 (2018)
93. P. Shine, M.D. Murphy, J. Upton, T. Scully, Machine-learning algorithms for predicting on-farm direct water and electricity consumption on pasture based dairy farms. Comput. Electron. Agric. **150**, 74–87 (2018)
94. M. Çunkaş, A. Altun, Long term electricity demand forecasting in Turkey using artificial neural networks. Energy Sources Part B Econ. Plan. Policy **5**(3), 279–289 (2010)
95. M. Kankal, E. Uzlu, Neural network approach with teaching–learning-based optimization for modeling and forecasting long-term electric energy demand in Turkey. Neural Comput. Appl. **28**(1), 737–747 (2017)
96. D. Singhal, K. Swarup, Electricity price forecasting using artificial neural networks. Int. J. Electr. Power Energy Syst. **33**(3), 550–555 (2011)

97. Ö.Ö. Bozkurt, G. Biricik, Z.C. Tayşi, Artificial neural network and SARIMA based models for power load forecasting in Turkish electricity market. PLoS One **12**(4), e0175915 (2017)
98. P. Jiang, R. Li, N. Liu, Y. Gao, A novel composite electricity demand forecasting framework by data processing and optimized support vector machine. Appl. Energy **260**, 114243 (2020)
99. S. Maldonado, A. Gonzalez, S. Crone, Automatic time series analysis for electric load forecasting via support vector regression. Appl. Soft Comput. **83**, 105616 (2019)
100. Y. Chen et al., Short-term electrical load forecasting using the Support Vector Regression (SVR) model to calculate the demand response baseline for office buildings. Appl. Energy **195**, 659–670 (2017)
101. F. Kaytez, A hybrid approach based on autoregressive integrated moving average and least-square support vector machine for long-term forecasting of net electricity consumption. Energy **197**, 117200 (2020)
102. M. Zahid et al., Electricity price and load forecasting using enhanced convolutional neural network and enhanced support vector regression in smart grids. Electronics **8**(2), 122 (2019)
103. G.-F. Fan, Y.-H. Guo, J.-M. Zheng, W.-C. Hong, Application of the weighted k-nearest neighbor algorithm for short-term load forecasting. Energies **12**(5), 916 (2019)
104. B. Sun, W. Cheng, P. Goswami, G. Bai, Short-term traffic forecasting using self-adjusting k-nearest neighbours. IET Intell. Transp. Syst. **12**(1), 41–48 (2018)
105. D.H. Wolpert, Stacked generalization. Neural Netw. **5**(2), 241–259 (1992). https://doi.org/10.1016/S0893-6080(05)80023-1
106. J. Mei, D. He, R. Harley, T. Habetler, G. Qu, A random forest method for real-time price forecasting in New York electricity market, in *2014 IEEE PES General Meeting | Conference Exposition*, (2014), pp. 1–5. https://doi.org/10.1109/PESGM.2014.6939932
107. J. Moon, Y. Kim, M. Son, E. Hwang, Hybrid short-term load forecasting scheme using Random Forest and multilayer perceptron. Energies **11**(12) (2018). https://doi.org/10.3390/en11123283
108. T. Ahmad, H. Chen, Nonlinear autoregressive and random forest approaches to forecasting electricity load for utility energy management systems. Sustain. Cities Soc. **45**, 460–473 (2019). https://doi.org/10.1016/j.scs.2018.12.013
109. M. Meng, C. Song, Daily photovoltaic power generation forecasting model based on Random Forest algorithm for North China in winter. Sustainability **12**(6) (2020). https://doi.org/10.3390/su12062247
110. P. Gaillard, Y. Goude, Forecasting electricity consumption by aggregating experts; how to design a good set of experts, in *Modeling and Stochastic Learning for Forecasting in High Dimensions*, (Springer, Cham, 2015), pp. 95–115
111. V. Mayrink, H.S. Hippert, A hybrid method using exponential smoothing and gradient boosting for electrical short-term load forecasting, in *2016 IEEE Latin American Conference on Computational Intelligence (LA-CCI)*, (2016), pp. 1–6. https://doi.org/10.1109/LA-CCI.2016.7885697
112. H. Zheng, J. Yuan, L. Chen, Short-term load forecasting using EMD-LSTM neural networks with a Xgboost algorithm for feature importance evaluation. Energies **10**(8) (2017). https://doi.org/10.3390/en10081168
113. T. Pinto, I. Praça, Z. Vale, J. Silva, Ensemble learning for electricity consumption forecasting in office buildings. Neurocomputing **423**, 747–755 (2021). https://doi.org/10.1016/j.neucom.2020.02.124
114. L. Wang, S.-X. Lv, Y.-R. Zeng, Effective sparse adaboost method with ESN and FOA for industrial electricity consumption forecasting in China. Energy **155**, 1013–1031 (2018). https://doi.org/10.1016/j.energy.2018.04.175
115. R.K. Agrawal, F. Muchahary, M.M. Tripathi, Ensemble of relevance vector machines and boosted trees for electricity price forecasting. Appl. Energy **250**, 540–548 (2019). https://doi.org/10.1016/j.apenergy.2019.05.062

116. F. Zhang, H. Fleyeh, C. Bales, A hybrid model based on bidirectional long short-term memory neural network and Catboost for short-term electricity spot price forecasting. J. Oper. Res. Soc. **0**(0), 1–25 (2020). https://doi.org/10.1080/01605682.2020.1843976
117. L. Deng, D. Yu, Deep learning: Methods and applications. Found Trends Signal Process **7**(3–4), 197–387 (2014). https://doi.org/10.1561/2000000039
118. I. Goodfellow, Y. Bengio, A. Courville, *Deep Learning* (MIT Press, 2016). [Online]. Available: https://books.google.com.tr/books?id=omivDQAAQBAJ
119. S. Ferlito et al., Predictive models for building's energy consumption: An Artificial Neural Network (ANN) approach, in *2015 xviii aisem annual conference*, (2015), pp. 1–4
120. J. Catalão, S. Mariano, V. Mendes, L. Ferreira, An artificial neural network approach for short-term electricity prices forecasting, in *2007 International Conference on Intelligent Systems Applications to Power Systems*, (2007), pp. 1–6
121. D.E. Rumelhart, G.E. Hinton, R.J. Williams, *Learning Internal Representations by Error Propagation* (California Univ San Diego La Jolla Inst for Cognitive Science, 1985)
122. P.J. Werbos, Generalization of backpropagation with application to a recurrent gas market model. Neural Netw. **1**(4), 339–356 (1988)
123. R.J. Williams, D. Zipser, Gradient-based learning algorithms for recurrent. Backpropagation Theory Archit. Appl. **433**, 17 (1995)
124. J.L. Elman, Finding structure in time. Cogn. Sci. **14**(2), 179–211 (1990)
125. M.I. Jordan, Serial order: A parallel distributed processing approach, in *Advances in Psychology*, vol. 121, (Elsevier, 1997), pp. 471–495
126. M. Schuster, K.K. Paliwal, Bidirectional recurrent neural networks. IEEE Trans. Signal Process. **45**(11), 2673–2681 (1997)
127. K. Cho et al., Learning phrase representations using RNN encoder-decoder for statistical machine translation, *ArXiv Prepr. ArXiv14061078*, 2014
128. J. Chung, C. Gulcehre, K. Cho, Y. Bengio, Empirical evaluation of gated recurrent neural networks on sequence modeling, *ArXiv Prepr. ArXiv14123555*, (2014)
129. S. Hochreiter, J. Schmidhuber, Long short-term memory. Neural Comput. **9**(8), 1735–1780 (1997)
130. R. Dey, F.M. Salem, Gate-variants of gated recurrent unit (GRU) neural networks, in *2017 IEEE 60th International Midwest Symposium on Circuits and Systems (MWSCAS)*, (2017), pp. 1597–1600
131. J. Schmidhuber, Deep learning in neural networks: An overview. Neural Netw. **61**, 85–117 (2015)
132. Y. Wang, Y. Liu, M. Wang, R. Liu, Lstm model optimization on stock price forecasting, in *2018 17th International Symposium on Distributed Computing and Applications for Business Engineering and Science (DCABES)*, (2018), pp. 173–177
133. L. Sehovac, C. Nesen, K. Grolinger, Forecasting building energy consumption with deep learning: A sequence to sequence approach, in *2019 IEEE International Congress on Internet of Things (ICIOT)*, (2019), pp. 108–116
134. D.L. Marino, K. Amarasinghe, M. Manic, Building energy load forecasting using deep neural networks, in *IECON 2016-42nd Annual Conference of the IEEE Industrial Electronics Society*, (2016), pp. 7046–7051
135. S. Ayvaz, O. Arslan, Forecasting electricity consumption using deep learning methods with hyperparameter tuning, in *2020 28th Signal Processing and Communications Applications Conference (SIU)*, (2020), pp. 1–4
136. R.A. Horn, The hadamard product. Proc. Symp. Appl. Math. **40**, 87–169 (1990)
137. R.A. Horn, C.R. Johnson, *Matrix Analysis* (Cambridge University Press, 2012)
138. G. Xiuyun, W. Ying, G. Yang, S. Chengzhi, X. Wen, Y. Yimiao, Short-term load forecasting model of gru network based on deep learning framework, in *2018 2nd IEEE Conference on Energy Internet and Energy System Integration (EI2)*, (2018), pp. 1–4
139. M. Sajjad et al., Towards efficient building designing: Heating and cooling load prediction via multi-output model. Sensors **20**(22), 6419 (2020)

140. M. Sajjad et al., A novel CNN-GRU-based hybrid approach for short-term residential load forecasting. IEEE Access **8**, 143759–143768 (2020)
141. J. Snoek, H. Larochelle, R.P. Adams, Practical Bayesian optimization of machine learning algorithms, in *Advances in Neural Information Processing Systems*, vol. 25, (2012). [Online]. Available: https://proceedings.neurips.cc/paper/2012/file/05311655a15b75fab86956663e1819cd-Paper.pdf
142. Z. Chang, Y. Zhang, W. Chen, Effective Adam-optimized LSTM neural network for electricity price forecasting, in *2018 IEEE 9th International Conference on Software Engineering and Service Science (ICSESS)*, (2018), pp. 245–248
143. R.J. Hyndman, A.B. Koehler, Another look at measures of forecast accuracy. Int. J. Forecast. **22**(4), 679–688 (2006)
144. T. Chai, R.R. Draxler, Root mean square error (RMSE) or mean absolute error (MAE)?–Arguments against avoiding RMSE in the literature. Geosci. Model Dev. **7**(3), 1247–1250 (2014)
145. C.J. Willmott, K. Matsuura, Advantages of the mean absolute error (MAE) over the root mean square error (RMSE) in assessing average model performance. Clim. Res. **30**(1), 79–82 (2005)
146. A. De Myttenaere, B. Golden, B. Le Grand, F. Rossi, Mean absolute percentage error for regression models. Neurocomputing **192**, 38–48 (2016)

Index

M. Rahmani-Andebili (ed.), *Applications of Artificial Intelligence in Planning and Operation of Smart Grids*, Power Systems,
https://doi.org/10.1007/978-3-030-94522-0

GPSR Compliance
The European Union's (EU) General Product Safety Regulation (GPSR) is a set of rules that requires consumer products to be safe and our obligations to ensure this.

If you have any concerns about our products, you can contact us on

ProductSafety@springernature.com

In case Publisher is established outside the EU, the EU authorized representative is:

Springer Nature Customer Service Center GmbH
Europaplatz 3
69115 Heidelberg, Germany

www.ingramcontent.com/pod-product-compliance
Ingram Content Group UK Ltd.
Pitfield, Milton Keynes, MK11 3LW, UK
UKHW021834270726
14058UKWH00001B/149

* 9 7 8 3 0 3 0 9 4 5 2 4 4 *